Contents

BRUTE FORCE

CRACKING THE DATA ENCRYPTION STANDARD

Matt Curtin

INTERHACK CORPORATION

Copernicus Books

AN IMPRINT OF SPRINGER SCIENCE+BUSINESS MEDIA

Published in the United States by Copernicus Books,
an imprint of Springer Science+Business Media.

Copernicus Books
Springer
233 Spring Street
New York, NY 10013

www.springeronline.com

Library of Congress Cataloging-in-Publication Data
Curtin, Matt.
Brute force : Cracking the data encryption standard / Matt Curtin.
p. cm.
Includes bibliographical references and index.
ISBN 0-387-20109-2 (alk. paper)
1. Computer security. 2. Data encryption (Computer science) I.Title.
QA76.9.A25C873 2005
005.8′2—dc22

2004058919

Manufactured in the United States of America.
Printed on acid-free paper.

9 8 7 6 5 4 3 2 1

ISBN 0-387-20109-2 SPIN 10958636

Foreword

A big battle over privacy was fought in the 1970s, 80s, and 90s, and most people didn't even know it was happening.

The U.S. government deliberately restricted the ways in which people could protect their own privacy. They did this with laws, with regulations, and by threatening prominent activists like Ron Rivest and Phil Zimmermann with censorship and prosecution. Most of it was unconstitutional, though they got away with it for decades. But most importantly, they restricted our privacy by keeping us ignorant and by lying to us.

A good way to keep information private is to safeguard it with encryption, a mathematical technology that scrambles information. You set it up so that the only people who have the "key" to unscramble it are the people that the owner intends to give access to. The government wanted to keep a monopoly on information about encryption. This would let the government hide information from its citizens (and from foreigners), but its own citizens (and foreigners) could not hide information from the government. The government had already threatened prominent academic researchers, tried to cut off National Science Foundation funding for research in encryption, and had built a "voluntary" censorship system for research papers.

It seemed to some people that freedom to do research, freedom to publish the results, and privacy were fundamental values of society that were more important than any particular government desires. The early academic researchers of cryptography, like David Chaum, Ron Rivest, and Whitfield Diffie, were such people. The Cypherpunks, who came along a few decades later, were also such people. I co-founded the Cypherpunks, an open group who educated ourselves and each other about encryption, and encouraged each other to write encryption software for free public use. Our goal was to re-establish the freedoms that the government had silently taken away, do the research, and publish the results, to transform society's expectations about privacy.

Part of the lies and ignorance created by the government was about a system called DES—the Data Encryption Standard. The government claimed that it was secure and private. Independent researchers claimed that it was too easy for governments to break into the privacy of DES. But mere claims were not enough to stop it, and the government succeeded in getting almost everyone to use DES worldwide. Banks used it to secure billions of dollars of money transfers. Satellite TV companies used it to keep their transmissions to their customers private. Computer security products used it. ATMs used it to guard the phone line that connects them to their bank and tells them when to deliver cash.

DES was deliberately designed by the U.S. government to be flawed. The government could read what was encrypted by DES, merely by spending enough money to build a machine that would break it. And the amount of money that it took went down every year, both as technology evolved, and as the designer learned more about how to build such machines. All that knowledge was hidden in the same secretive government agencies who deliberately weakened DES.

As personal computers and chip technology rapidly became cheaper and faster, ordinary people working together could rival the machine-building power of the government. This book is the story of how they proved the government was lying, twenty years after the lie, and by doing so, energized the public to take its privacy into its own hands. The end result was not only that government policy about encryption and privacy was changed. Also, the process of building networks of people and machines to do calculations by "brute force" taught us a lot about collaboration, about social structures in volunteer groups, about how the world is changed by the broad distribution of consumer products that compute. And about how to break down certain kinds of intractable problems into small pieces, such that many people can do a piece and thus contribute to the solution.

The panicky public reaction to the attack of 9/11 was unable to upset the balance of relatively sane encryption policy that it had taken decades to set right. However, the abdication of responsibility that took hold of both the Congress and the bulk of the public has let a corrupt administration get away with murder—literally, in the case of hundreds of thousands of civilians in Iraq. Civil rights and moral standards as basic as the prohibition on torture, the freedom to move around in your own country, and the universal condemnation of unprovoked attacks on other countries have all fallen by the wayside.

Yet computers and networks have shown even more interesting ways for millions of people to collaborate to solve big intractable problems like this. As I write this, thousands of people are working for a few days from their homes, phoning up strangers to encourage them to go out and vote in the upcoming U.S. election. A computer network, programmed by a small number of people, has collected and connected both the callers and the people who they should call.

We will continue to be surprised by the capabilities that human societies have, when thousands of people network through their computers to accomplish a common purpose.

John Gilmore
Electronic Frontier Foundation
October 31, 2004

Preface

In the past fifty years, society has undergone a radical shift in the storage and processing of information, away from the physical and toward electronic representation. Important information is no longer written on a sheet of paper and stored in a locked file cabinet or safe. Information necessary to care for our health, our finances, and the institutions, public and private, that support society is now stored electronically, in little ones and zeroes. Encryption technology—the mathematical system used to protect electronic information—was developed to protect all of those data from prying eyes.

In the late 1970s, the U.S. government decided to create a national data encryption standard in order to bring order to a market that had generated a multitude of competing and rarely complimentary encryption products. The standard the government settled on called the data encryption standard or DES was immediately criticized for being too weak by many security and computer experts. For years the critics demanded stronger cryptography and for years the government ignored their requests.

In 1997 a security company, RSA, answered DES's critics. They launched a contest, challenging cryptographers and computer enthusiasts to show the government just how weak DES was. *Brute Force* tells the story of DES: how it was established, challenged, and ultimately defeated. But more than the longevity of DES or the definition of the standard was at stake.

Even while technologists argued over how strong the cryptographic standard had to be, lawmakers in the United States were busy debating the government's role in the regulation of cryptography. At the heart of the debate was whether or not the government would permit American companies to export products that they couldn't break overseas, and whether private citizens would be permitted to use cryptography that would shield their information from the eyes of government. Libertarians, cryptographers, and security experts wanted to be able to use and export the most robust encryption possible. While some in Congress supported this view, many other members of the government, including the Clinton administration, were wary of strong encryption, fearing it would fall into the hands of criminals and terrorists. *Brute Force* tells the story of the legislative battle over DES as well.

Although cryptographic specialists will likely be familiar with parts of this story and be eager to learn what happened behind the scenes, this is not only a story for technologists. What happened in 1997 affects people everywhere, even today, and will do so for years to come. So long as we store and transmit private information on computers, we will need to protect it from those who would try to steal it.

Events of this story fall into one of three major topics: the technology of secret writing, the story of how people who never knew each other came together to defeat the global standard for secret writing, and the wrangling over public policy on cryptography. The story is told not by recounting events in a strictly chronological order but as chains of events that place different parts of the story into context and allow the reader to see how these events finally came crashing together, changing the face of information management forever.

Acknowledgments

This book is the product of tremendous work by many people. Thanks must go to Peter Trei for suggesting the demonstration of a brute force attack on the Data Encryption Standard and to RSA for sponsoring the contest that at long last demonstrated the weakness of DES. I also offer my heartfelt thanks to Rocke Verser for his work in starting and running the DESCHALL project that participated in RSA's contest. Justin Dolske, Karl Runge, and the rest of the DESCHALL developers also put in many hours to ensure our project's success and were as pleasant and interesting as one could hope for. Not to be forgotten are the thousands of people who participated by running the DESCHALL client programs on their computers, telling their friends about our project, and giving us access to the tremendous computational power needed to verify that strong cryptography makes the world a safer place. Telling the story of this significant period in the history of cryptography in the form of the book that you are now holding proved to become another sizable project. Gary Cornell at Apress got me connected with the right people at Copernicus Books. I appreciate the connection as well as the help that Anna Painter, Paul Farrell, and the rest of the folks at Copernicus Books provided in moving the book from a raw manuscript into its final, published form. Thanks are also due to John Gilmore for resurrecting a recording of Martin Hellman and Whitfield Diffie arguing with government representatives the need for a stronger standard than what became codified in DES. The recording and other electronic resources of interest are available at:

http://ergo-sum.us/brute-force/.

Finally I thank my wife Nicole for her continued support and thoughtful interest in my work.

Matt Curtin
December 2004

To the Cypherpunks—
making the networks safe for privacy...

1

Working Late

June 17, 1997, 11:51 P.M.
Salt Lake City, Utah

A modest desktop computer quietly hummed along. It sat in the offices of iNetZ Corporation, a Web services company started just a few months earlier. This machine, just an ordinary machine with a 90 MHz Intel Pentium processor, was still hard at work in the darkness of an office that had closed for the day several hours earlier. Running a program called DESCHALL—pronounced "DESS-chall" by some, and "dess-SHALL" by others—this computer was trying to read a secret message. After all, it was practically the middle of the night, and the machine had nothing else to do.

The secret message was protected by the U.S. government standard for data encryption, DES. Largely as a result of the government's fiat, DES was used to protect sensitive data stored on computers in banking, insurance, health care, and essentially every other industry in nearly every part of the world. It was a U.S. standard, but in a world of international corporations and global trade increasingly conducted by computer, it was in everyone's interest, or so it seemed, to standardize on DES.

The slowest of eight iNetZ machines on which system administrator Michael K. Sanders installed DESCHALL, the quiet little computer was trying to find the single key out of more than 72 quadrillion (72,000,000,000,000,000) that would unlock the secret message. Applying one key after another to the message and checking the output for something intelligible, the machine was trying some 250,000 keys per second. It did not falter. It did not quit. It just kept banging away at the problem.

Quite suddenly, just before midnight, the computer's DESCHALL program came to a halt.

When Sanders came to work at iNetZ the following morning, this unassuming computer was displaying an urgent message on its screen.

Information security would never be the same.

2

Keeping Secrets

Cryptography is quite simply the practice of secret writing. The word itself comes from two Greek words, *kryptos* ("hidden") and *graphein* ("writing"). With a history going back at least 4000 years, cryptography has long been surrounded by mystery and intrigue.

Ancient Egyptians used cryptography in hieroglyphic writing on some monuments, thus protecting some proper names and titles. Some 2000 years ago, Julius Caesar used a simple system of substituting one letter for another to send secret messages to his generals. In the thirteenth century, English mathematician Roger Bacon wrote of systems to write in secret in his "Concerning the Marvelous Power of Art and of Nature and Concerning the Nullity of Magic." In that document, Bacon enumerated seven methods for secret writing and famously opined, "A man who writes a secret is crazy unless he conceals it from the crowd and leaves it so that it can be understood only by effort of the studious and wise."

Throughout its history, cryptography has primarily been a tool of government elites because they were the ultimate keepers of military and diplomatic secrets. Code makers and breakers alike have thus almost always been employed by governments to discover others' secrets while protecting their own.

Cryptography is important because it enables information to be stored and transmitted secretly. The ability to control the flow of information, to enforce who may and may not know a particular fact is precisely the kind of power that traditionally governments and increasingly private businesses seek to wield against adversaries and competitors. Especially when the keepers of a secret are not able to meet together, out of the range of eavesdroppers and spies, there is a need for communicating secretly right in the open. As had been demonstrated in numerous wars of the twentieth century, anyone can intercept radio signals. Telephone lines can be tapped. This is where cryptography comes into play—locking up information so that it will remain secret while it is being transmitted via a medium that is open to all.

Once we had passed the age of the trusted courier and locked box, new telegraph and especially radio technologies created the need for reliable encryption machines. In the early twentieth century, enterprising inventors saw an opportunity and before 1920 had invested four such devices. At the heart of these machines was a series of three or four rotors—wired code wheels, each with twenty-six different electrical contacts on each side. To encrypt a message, the user would type a letter on the keyboard, such as A, and electrical current would flow through the machine, going through the rotors, and printing a completely different letter, such as V. The rightmost code wheel would then advance one position, and the user pressing A again would result in another letter being printed, such as T, before the code wheel rotated again. Once the rotor went through all twenty-six positions, the rotor next to it would also advance, much like an analog odometer on an automobile.

In this way, the user would type the original message, while the machine would produce *ciphertext* that could safely be sent as a radio signal. The intended recipient of the message would have a matching cipher machine that would turn the signal back into human-readable *plaintext*. In the United States, Edward H. Hebern invented his machine in 1917, Germany's Arthur Scherbius invented his in 1918, and 1919 saw the invention of a machine in the Netherlands by Alexander Koch and in Sweden by Arvid Gerhard Damm. Scherbius called his machine Enigma, and it would become the only financially successful cipher machine from the era.

Enigma was patented by Scherbius, an electrical engineer, and E. Richard Ritter, a certified engineer. After the eventual transfer of patent rights, Engima would come to be marketed commercially by *Chiffriermaschinen Aktien-Gesellschaft* (Cipher Machines Stock Corporation), whose board of directors included Scherbius and Ritter. Several governments began to investigate Engima, with variations of the original design eventually coming into use throughout the German, Italian, and Japanese armed forces.

Despite the best efforts of its producers, Engima was not generally accepted in the world of business. Its commercial success came as a result of the Axis use of the machine to protect military and diplomatic communications.[1]

With the rise of radio technology in government and military communications in the early twentieth century, the danger of messages being intercepted increased dramatically. Instead of having to get physical access to communications circuits such as telephone or telegraph lines, operatives could simply point high-powered antennas toward their targets and start listening. Governments throughout the world developed "signals intelligence" groups, chartered to intercept radio communications sent by other nations, and to report their findings to their own leaders. To protect their own communications from foreign signals intelligence efforts, governments began to encrypt their radio signals.

Governments would not easily give up the ability to read others' messages. Signal intelligence came to mean not just message interception but also breaking the encryption used to protect the messages. In the years leading up to World War II, the United States maintained an active signal intelligence operation even while hoping to avoid being drawn into the global conflict. In 1938, the Japanese empire began to use a machine they called "Alphabetical Typewriter 97" for their diplomatic messages—a rotor machine like Germany's Enigma. Unable to read those messages, the U.S. Army Signals Intelligence Service (SIS) began a project to break the Japanese system, which they had code-named, "Purple."

In the late 1930s, SIS cryptanalysts (code breakers) under the direction of cryptographic pioneer Frank Rowlett spent eighteen months studying intercepted Japanese diplomatic messages, looking for any clue that would help them to unlock Purple's secrets. One day in September 1940, SIS cryptanalyst Genevieve Grotjan made a critical discovery. She found important and previously undiscovered correlations among different messages encrypted with Purple. After Grotjan brought her discovery to the attention of the rest of the SIS Purple team, they were able to build a duplicate of a machine they had never seen—the Alphabetic Typewriter 97.[2]

Putting its new machine to work right away, SIS discovered that Purple was used not simply for routine traffic, but the most sensitive of the Japanese empire's secrets. Intelligence gathered from intercepted and decrypted Purple messages was so valuable that those decrypted intercepts came to be called "Magic" within SIS.

When Rowlett returned to his office from a meeting at midday on December 3, 1941, he picked up a Magic decrypt from his in-box. That message, intercepted just that morning, was directed to Japan's embassy in Washington. Rowlett read the bizarre orders for Japanese diplomats to destroy their code books and even one of the two Purple machines they had. Without their code books and with only one working Purple machine, the Japanese embassy simply would not be able to operate normally. Colonel Otis Stadtler, who was responsible for distributing Magic decrypts arrived as Rowlett was reading the message. After some discussion, Stadtler realized the meaning of the order: Japan was preparing to go to war with the United States.

On the evening of December 6, U.S. president Franklin D. Roosevelt received analysis of the intelligence: war with Japan was inevitable, and the Magic decrypts were used to support the conclusion. As the Japanese military used different codes from the Japanese diplomats, President Roosevelt had no way of knowing that on the very next day, Japan would attack Pearl Harbor and kill over 2300 Americans. Only five years later would there be enough time for SIS cryptanalysts to look at the military intercepts in the months before the strike on Pearl Harbor. Their efforts to break those messages proved successful, and they anguished over the results of their work. Though not naming Pearl Harbor explicitly, the Japanese military had been ordered to be on a footing for war with the United States by November 20, 1941.[3]

Private industry, driving much of the revolution in communication technology of the twentieth century, also developed its interest and expertise in cryptography. Claude E. Shannon at AT&T Bell Telephone Laboratories made several critical contributions to modern communication, computing, and cryptography. Shannon joined Bell Labs in 1941, after completing his Ph.D. in mathematics at the Massachusetts Institute of Technology. At Bell Labs, Shannon worked as a research mathematician and came to be known for "keeping to himself by day and riding his unicycle down the halls at night."[4]

In 1948, Shannon published "A Mathematical Theory of Communication" in the *Bell System Technical Journal*.[5] The paper was a breakthrough, founding the study of information theory, and coining

Fig. 1. Claude E. Shannon, c. 1952. Property of AT&T Archives. Reprinted with permission of AT&T.

the term "bit" to describe a BInary uniT. Up to that time, communication was thought to require electromagnetic waves down a wire or radio waves toward a receiver, but Shannon showed how words, pictures, and sounds could be sent across any medium that would carry a stream of bits. The following year, Shannon applied his work directly to cryptography in a paper entitled, "Communication Theory of Secrecy Systems."[6] This paper founded modern mathematically-based cryptography outside of government intelligence agencies.

The rise of the computer and the rise of cryptography have gone hand in hand. Computing technology has made exchanging information easier, making communication and collaboration easier. Since people still want—and in an ever-growing number of cases, are legally obligated—to stay in control of information in their stewardship, people need cryptography.

Code makers and code breakers agree: the computer is both friend and enemy. For cryptographers, computer technology makes the implementation and use of flexible cryptography easier, while frustrating the efforts of cryptanalysts. For cryptanalysts, the computer improves efficiency in the analysis of encrypted messages and building systems to undermine cryptography, thus making it easier to exploit any flaw in the cryptographers' creations.

Cryptosystems before the twentieth century required tedious manual processing of messages, using code books to match what was written to what was to be communicated, or perhaps a great deal of scratch paper to perform the necessary text substitution and transposition. The process of encrypting and decrypting messages essentially consisted of taking a handwritten message, looking up the correct corresponding symbol on a chart, and writing the symbol on the paper that would actually be delivered to the recipient, who would in turn look at the chart and convert the ciphertext back to the plaintext by hand, one letter at a time.

Later systems like Enigma, though more convenient than the "old way," were still cumbersome and slow. (Early Enigma promotion material boasted that the machine could process 300 characters per minute.)

Though the internal mechanics were much more complicated, the user of the Enigma might liken its operation to a typewriter where the keys are randomly reassigned. The sender would type the letter according to the keys written on the keyboard, knowing that when an `A` is struck, a `V`, for example, will be written. The recipient will then need to know the keyboard layout used by the sender in order to recognize that the `V` in the message was created by striking the `A` key, and write "A" on a scratch pad. Working letter by letter, the sender's message becomes visible. Enigma handled this substitution work automatically, preventing operators from needing scratch paper.

Now, with computers, recipients can often click a few buttons and have huge amounts of deciphered information almost instantly turned into the sender's original message.

Perhaps no one understood the challenge and opportunity that emerged in the post-war era better than the researchers at IBM. In the 1950s and 1960s, with its systems designed to handle the heaviest information processing needs of both corporations and government agencies, IBM had to give serious consideration to the handling of sensitive data.

One of the earliest applications for computers was in the handling of government information—some of which was protected by law. Security was just as much a requirement for early computer systems as the ability to store and to process information accurately.

The trend to establish standards for data security in automated information systems became an important issue for IBM and its customers. The possibility of computerized records being abused was not lost on Americans, who were fascinated with computers and technology, but also worried about the implications of their use in society. One of the key figures in helping IBM realize a workable, powerful security scheme was a German émigré by the name of Horst Feistel. Feistel had arrived in the United States decades earlier, in 1934. Despite his interest in cryptography, he avoided working in the field during World War II to avoid suspicion by the American authorities.

After the war, Feistel found employment at the U.S. Air Force Cambridge Research Center, where he worked on identify friend-or-foe (IFF) systems. IFF systems were (and still are) used on the battlefield to avoid "friendly fire" incidents, where forces attack allied units instead of the enemy. Radar systems with IFF capability, for example, report not only the position of units in range, but whether they are friendly or hostile—thanks to the use of cryptography.

In the middle of the twentieth century, the highly secretive U.S. National Security Agency (NSA) had a virtual monopoly on cryptographic research and were trying hard to maintain it. Feistel's Air Force project was canceled—though details are shrouded in military secrecy, NSA is generally credited with ensuring its hasty demise.

Feistel attempted to continue his work at Mitre Corporation in the 1960s, but again ran afoul of NSA's plans. Dependent on Department of Defense contracts, Mitre had little choice but to ask Feistel to direct his energies elsewhere—presumably also at NSA's behest.

Determined to apply his hard-earned expertise in cryptography, Feistel joined IBM before 1970, where he was finally free to continue his work, and headed up a research project known as Lucifer. The goal of Lucifer was to develop cryptographic systems for use in commercial products that would address the growing need for data security. IBM consequently was able to offer clients a means of protecting data stored in its computers.

Commercial users of computers were finally seeing the need to protect electronic information in their care, and an explosion began in the commercial availability of cryptographic products. In the late 1960s, fewer than five companies were offering cryptographic products, but by the early 1970s, more than 150 companies were active in the marketplace—and more than fifty of them were from outside of the U.S.

During this time, Feistel published an article in *Scientific American*, describing cryptography and how it relates to protecting private information in computers. Although much of the article focused on cipher machines of the sort that were used in World War II, it also contained some descriptions for mechanisms for computer software to encrypt information. Those methods, known as Feistel Networks, are the basis of many cryptosystems today.

Because the government kept their cryptographic technology under lock and key, commercial cryptographers could only guess at what their counterparts within government research facilities like NSA had achieved. These commercial cryptographers began with the fragments that could be assembled from historical literature and began to lay the foundation for open (i.e., not secret) cryptologic research.

At this time, though, very little was understood about how well various cryptographic techniques could withstand analysis. For example, one might believe that an encrypted message would be twice as resistant to analysis if encrypted twice. Only after years of research did cryptographers come to realize that for many kinds of ciphers, double encryption is no stronger than single encryption. Many questions played into a system's strength. How strong would a rotor-based system be if it used four rotors instead of three? How strong is strong enough? How strong is a rotor-based machine system by comparison with an encryption system implemented entirely in software?

In the early 1970s, no one outside of government cryptology knew the answers to questions like these, and it would be years before sufficient work in the field would be done to find answers. Thus, the availability of cryptographic products was of little help—people simply didn't know how good any of it was, and making meaningful comparisons was impossible. Even worse, no two vendors could agree on a system, requiring that both sender and receiver use the same equipment. It would be like buying a Ford only to discover that the nearest gas station sold only fuel to work with Chrysler cars.

Knowing that information needed to be protected, computer system managers had little choice but to buy *something* and hope for the best.

3

Data Encryption Standard

In the United States, the National Bureau of Standards (NBS) began undertaking an effort aimed at protecting communications data. As part of the Department of Commerce, NBS had an interest in ensuring that both its own systems and those of the commercial entities with which it dealt were adequately protecting the information under their stewardship.

The NBS effort included the establishment of a single standard for data encryption, which would allow products to be tested and certified for compliance. The establishment of a single standard would solve three major problems in the chaotic encryption marketplace. First, products compliant with the standard would have to meet security specifications established by experts in cryptography; individual amateurish efforts at merely obfuscating information would not pass muster. Second, compliant products from different vendors would be able to work with one another, allowing senders and recipients to buy from the vendors of their choosing. And third, the tremendous costs incurred by vendors in the creation of cryptographic systems could be reduced, since they would be able to focus on making the systems convenient to use, rather than spending huge amounts of money on development of the cryptographic underpinnings.

Requirements for the standard cryptographic algorithm—the definition of the series of steps needed to turn plaintext into ciphertext and back again—were published in the *Federal Register*. Among the requirements were a high level of security, complete and open specification, flexibility to support many different kinds of applications, efficiency, and exportability to the global marketplace.

NBS received many responses, though it ultimately determined that none of the algorithms submitted satisfied all of these requirements. Despite this apparent setback, NBS did not consider the effort to be a complete loss since it demonstrated that there was a substantial interest in cryptography outside of military circles. The large number of responses, in and of itself, was taken as a firm and positive step in the right direction.

NBS published a second request in the *Federal Register* on August 27, 1974. Once again, several serious submissions were made. Some were too specialized for the purposes NBS envisioned. Others were ineffective. One, however, showed great potential.

IBM's Lucifer project had an algorithm simply named "Lucifer," that was already in the latter stages of its development. IBM submitted a variation of the algorithm, one with a 112-bit key, to NBS.

Before the significance of the 112-bit key can be fully appreciated, it is important to note that modern computers are binary. That is, they store and process data in bits, the binary units Claude E. Shannon described in 1948. Anything with two settings can be used to represent bits. Consider a light bulb. It has two settings and two settings only: on and off.

All data in binary computers are represented in terms of bits, which are represented as 0 or 1. Absolutely everything, to be stored into a computer, must ultimately be represented with these two, and only these two, digits.

The easiest way to grasp the security of algorithms like IBM's Lucifer is to imagine a simple bicycle tumbler lock. Usually, such locks are made up of four or five tumblers, each with ten positions, labeled 0 through 9. In digital computers, however, a cryptosystem with a 112-bit key is like having a lock with 112 tumblers, each with two settings, `0` and `1`.

IBM's algorithm therefore had a total of 2^{112} possible settings, only one of which was the "key" to the system, the equivalent of the setting of a bicycle lock which would allow its opening. Seeing that number written out—5,192,296,858,534,827,628,530,496,329,220,096—shows why scientists prefer to use exponents when talking about large numbers. The difference is even more pronounced (pardon the pun) when you hear the numbers spoken. "One hundred twelve bit" is much easier to say than "five decillion one hundred ninety-two nonillion two hundred ninety-six octillion eight hundred fifty-eight septillion five hundred thirty-four sextillion eight hundred twenty-seven quintillion six hundred twenty-eight quadrillion five hundred thirty trillion four hundred ninety-six billion three hundred twenty-nine million two hundred twenty thousand ninety-six." Such a vast number of possible solutions made the Lucifer algorithm a powerful means to protect information—satisfying two important NBS criteria at once: high security and security coming from the key.

NBS saw IBM's submission as promising, but it had a serious problem—the algorithm was covered by some IBM patents which ruled out interoperability. IBM agreed to work out rights for the patents, such that even competitors would have the ability to produce systems that implemented the algorithm without the need to pay IBM licensing fees. Once this legal obstacle was removed, NBS went to work on evaluation of the system itself.

Lacking a staff with its own cryptographic expertise, NBS turned to the greatest source of cryptographic expertise known to exist—in other words, NSA—for help in evaluating the strength of the Lucifer algorithm. After careful analysis, NSA proposed two significant changes.

The first was a change in the algorithm's S-boxes. S-boxes are the part of the algorithm that control how the data are permutated as they move from step to step along the process of being converted from the readable message to the encrypted result (or vice-versa), much like the rotors of Enigma.

The second, and more drastic, was the reduction of key length from 112 to 56 bits. This recommendation came as a result of debate inside of NSA. While the code-making part of NSA wanted to produce a standard that was strong and could protect U.S. interests, the code-breaking part of NSA was concerned that if the standard were too strong, it could be used by foreign governments to undermine NSA's foreign signal intelligence efforts. Ultimately, 56 bits was the key size that won out as those two concerns were balanced.[7]

The difference in key size is significant. Because we're talking about "tumblers" that are binary here—we're working with a base of 2. That means that each digit added to the key doubles the key strength. That is, the number of possible settings, only one of which is the key to unlocking the encrypted message. Consider Table 1.

Power	*Conventional Notation*
2^1	2
2^2	4
2^3	8
2^4	16
2^5	32
2^9	512
2^{56}	72,057,594,037,927,936
2^{112}	5,192,296,858,534,827,628,530,496,329,220,096
2^{128}	340,282,366,920,938,463,463,374,607,431,768,211,456

Table 1. Powers of Two

The key of IBM's original cipher would be not just double or triple the strength of NSA's modification, but *fifty-six times* the strength. The reduction of the key rate caused a significant stir among the nascent group of civilian cryptographers.

In 1975, two cryptographers from Stanford became particularly critical of the 56-bit key. Whitfield Diffie, one of the two cryptographers, took the notion of an independent cryptographer to a new level. Not only was Diffie free from the restraints of secret government research, but he also developed his work free of the influence of large corporations. Having graduated from MIT with a degree in mathematics in 1965 and performed computer security work for several companies since then, Diffie found himself becoming recognized as an expert by his peers even without the help of a powerful support system.

Cryptographic systems long had a serious problem: getting the keys sent between the sender and recipient of encrypted messages. After all, if you can safely send a key in secret, why not use the same method to send the message itself? In practice, this problem was addressed through procedures, such as having the sender and recipient agree on a series of keys in person. The first message would be encrypted with the first key, the second with the next key, and so on, until they had exhausted their supply of keys, at which point they would need again to exchange a list of keys—whether in person or through a trusted source like a secured courier.

Being fascinated with the problem of the distribution of cryptographic keys, in particular key distribution over a global Internet, Diffie spent a lot of time thinking about this problem. While still forming his ideas on key distribution, Diffie visited IBM's Thomas J. Watson Laboratory to deliver a talk on cryptography, with particular emphasis on how to manage keys safely.

After his presentation, he learned that Martin Hellman, a professor of electrical engineering from Stanford had spoken at the same laboratory on the same topic not long before. Diffie took particular interest in Hellman because most cryptographers at the time were enamored with the algorithms themselves, leaving few to give the problem of key distribution any serious consideration.

That evening, Diffie got into his car and started driving across the country to meet Hellman. After arriving in Stanford, Diffie called Hellman, who agreed to a meeting. The two were impressed enough with each other that they looked for a way to work together. Because Hellman did not have the funding to hire Diffie as a researcher, he took Diffie on as a graduate student instead. Thus began the partnership of Diffie and Hellman at Stanford University.[8]

After the criticisms Hellman and Diffie leveled against the 56-bit key of the developing standard for data encryption throughout 1975 were ignored by NBS, the Stanford pair authored a letter published in *Communications of the ACM*. In that letter, they outlined their objections to the key size and its ramifications. Because the Association for Computing Machinery (ACM) is the oldest and largest association of computer scientists and engineers, its *Communications* is well-read and highly-regarded, seen by effectively everyone working in computing at the time.

Hellman and Diffie knew that the help of this group would be critical in forcing NBS to address their concerns. Even so, they recognized that the issue of the algorithm's security would be so far-reaching that their concerns would be of interest to the American public. The algorithm would protect data about the medical histories, finances, and official records of Americans from all walks of life.

If the standard could not withstand attack, it would be the American people who would suffer. Recognizing the difficulty of bringing such an obscure (albeit important) matter to the attention of the public, Hellman and Diffie wisely enlisted the help of David Kahn, author of the highly regarded 1967 book *The Codebreakers.*[9] Kahn wrote an Op-Ed piece for *The New York Times* that was published on April 3, 1976. In that article, Kahn wrote of the proposed standard, "While this cipher has been made just strong enough to withstand commercial attempts to break it, it has been left just weak enough to yield to government cryptanalysis."

By this time, experts from IBM, Bell Labs, and MIT had also weighed in on the matter: 56-bit keys were too small, they all declared. As Kahn noted in his article, "one major New York bank has decided not to use the proposed cipher" in part because of the criticisms of its key size.

The uproar was sufficient to cause the U.S. House of Representatives' Government Information and Individual Rights Subcommittee to look into the matter. NBS was forced to recognize that the field of cryptanalysis existed beyond the walls of government, that the concerns are real, and they must be addressed if the effort to standardize the proposed 56-bit system was to succeed.[10]Consequently, NBS decided to hold two workshops on the cipher proposed as the "data encryption standard" (DES).

NBS held two workshops in 1976 to deal with the objections raised by Hellman and Diffie. These were working meetings where cryptographers from across the country would be able to discuss the thorny issues around the proposed data encryption standard face-to-face. As part of their objections, Hellman and Diffie proposed the design of a special-purpose computer that would use a technique called brute-force to crack DES-encoded keys quickly. The first NBS workshop was composed of hardware experts who considered the proposed special-purpose DES cracker.

Some participants argued that the proposed DES cracking machine would not work because design and control costs would exceed the cost of the hardware. Hellman and Diffie countered that cracking DES keys would not be one large job, but many small jobs that could be performed independently. As such, there was no need for the microprocessors—the "brains" of the computer—to interact with one another. Each could be given tasks to perform independent of the others. This, Hellman and Diffie responded, meant that the objection to the feasibility of a brute-force attack on the basis of design and control costs did not stand.

Another matter of concern was the reliability of the computer—a more visible concern in the computing technology of the 1970s than it is today. The reliability of computers is directly tied to the number of components needed to construct them. Some of the NBS workshop participants performed calculations for a DES cracker with 1 million components—parts for handling computer working memory, storage, central processing, arithmetic logic, and all of the electronics to hold it all together. Based on the average time it would take electronic equipment of the day to fail, the million-component machine would not be able to run for more than a single day before failing in some way. Such a large system, with that level of failure, would be too big and too complex to operate.

The Diffie-Hellman design for a DES cracker, however, called for far fewer components—only 16,000. Furthermore, rather than using a large number of parts that would be used only a few times in the machine, the Diffie-Hellman design called for construction involving fewer types of parts—allowing any parts that fail to be easily replaced, getting the system back up and running in under ten minutes. Such a system would give error-free operation with a relatively small number of spare parts.

Another objection on the million-chip machine was its size: 6000 large cases—known as "racks"—that were 6 feet high. Hellman and Diffie responded with a proposal for a million chip machine in only 64 racks, suggesting that even were 1 million chips necessary, the size of the machine was being seriously overestimated.

Still basing assumptions on the large, million-chip, 6000-rack machine, power requirements were the next objection raised by NBS and others. Simply providing the electricity for such a machine to run would exceed any "reasonable budget," apparently without specifying what would constitute "reasonable." Hellman and Diffie proposed the use of chips manufactured in a newer and more cost-effective manner that would bring the operating cost to under $1500 per day, observing that power costs could be reduced five times with newer technology.

Looking at the speed with which a message could be encrypted with DES on readily available (general-purpose) chips, some participants determined that those chips would be too slow and cost too much when purchased in the quantity needed to test DES keys quickly. Looking at available technology, Hellman and Diffie suggested that complaints about chip speed and cost could be overcome by using a special chip, designed for the specific purpose of searching for DES keys. A special-purpose chip would dramatically increase the speed of the operation. Such chips, they observed, could be produced in quantity for $10 each.

In the course of this dispute, NBS even offered some of its own alternatives to increasing the key size. One approach they suggested was to develop a system that made use of frequent key changes. Rather than reusing the same key from one message to another, such a system would give each message a unique key. That way, the illicit discovery of a key would compromise only one message, rather than every message encrypted with that machine. Hellman and Diffie responded by observing that rather than cracking the message immediately after it was sent, some attackers might have the ability to intercept a message and then to spend the time necessary to break any particular message. (Interestingly, while cryptographers like Hellman and Diffie had no way to know it at the time, this is precisely what happened when SIS cryptanalysts could not keep up with the flow of Japanese military communications in the run-up to the attack on Pearl Harbor. Recall that SIS decrypted those messages five years after they were intercepted.) Hellman and Diffie went on to observe that medical records needed to remain private for ten years—that kind of long-term privacy requirement could not be met by a system where a single message encrypted with a relatively small key could be broken in a ten-year period.

Looking at the costs that would need to be borne by anyone implementing commercial cryptography, some argued that increasing the proposed standard's length of a key to 128 or 256 bits—as Hellman and Diffie suggested—would greatly increase the costs. The expense, in turn, would make the construction and use of such systems less attractive while also decreasing the overall use of encryption. Hellman and Diffie countered these assertions by observing that the computing power needed to perform encryption is much less than needed to perform brute-force search. (This works much like a scavenger hunt. Hiding twenty items—akin to encryption—is not significantly harder than hiding ten items, though finding those twenty—akin to brute-force decryption—would take dramatically more time than finding ten.) The difference in the cost of operation of a 128-bit system and a 56-bit system was negligible, but the payoff in terms of greater security was significant.

Finally, NBS argued that there simply was no way to tell for sure when the right key had been found in a brute-force search, even if someone took an encrypted message and used that key to turn it into a readable plaintext. Hellman and Diffie argued that while a formal proof would be difficult, the design of DES was not such that a ciphertext message would be able to decrypt into lots of different sensible-looking plaintext messages. The decryption process would produce either a sensible message or gibberish.

Hellman and Diffie argued that none of the NBS objections was valid and that a 56-bit key could not provide adequate security against a dedicated attacker. They recommended devices that would support variable key lengths. Allowing users to choose their own key lengths would allow them to decide for themselves whether the extra security of the larger keys was worth the extra time needed for the encryption and decryption processes.

NBS didn't stop with consideration of DES-cracking computers. The following month, NBS held a second workshop on DES, focused on the mathematical foundations for the DES algorithm. Participants in the second workshop expressed significant concern that while the design was available for review, the principles that guided NSA's changes were classified, and therefore available only to government cryptographers sworn to secrecy. The workshop adjourned without consensus.

Nevertheless, the workshops had three important effects. First, much concern was voiced over the possible weaknesses of DES, with the key length being a major issue, as well as the participants' inability to review the design principles behind NSA's S-Box changes. If NSA wanted to implant a secret "shortcut" so that only it could decrypt messages immediately, that would be the place to do it, and participants might not have enough understanding of the details to identify it.

Second, few participants were convinced that the Hellman-Diffie scheme for breaking DES keys was practical. Costs still seemed too high, and effort needed still seemed too great to be worthwhile. Given the technology of 1976 and the next few years, there seemed little likelihood that DES would be defeated by brute force.

Third, the arguments put forth by Hellman and Diffie did convince participants that the key length provided no safety margin. Essentially, the Hellman-Diffie designs for key-cracking computers were possible, but not presently feasible. Anything that would change that balance, driving the cost of computing down in an unexpected way would undermine the strength of DES against brute-force attacks.

NBS considered the matter as resolved as it would ever be, ultimately ignoring the warnings issued by the outsiders from Stanford and effectively declaring no need for a safety margin.

Whitfield Diffie and Martin Hellman documented their objections to the 56-bit key of the DES cryptographic algorithm in an article published in the June 1977 issue of *IEEE Computer*. Their article, "Exhaustive Cryptanalysis of the NBS Data Encryption Standard," described a special-purpose machine to crack DES keys by brute force.

Building on top of the debates during the NBS DES standardization process over the hardware requirements for DES-key-cracking computers, the published Diffie and Hellman design was estimated to cost $20 million to build, and would be able to break DES keys in roughly twelve hours each.

Four and a half years after announcing its intention to create a standard for data encryption, NBS published its official standard in the Federal Information Processing Standard series, a group of regulations and standards that all of the agencies in the Federal government must follow. At long last, FIPS 46, titled "Data Encryption Standard," was released.[11]

A private, non-profit industry association, the American National Standards Institute (ANSI) had (and still has) a committee to handle the standardization of information technology. Not wanting to duplicate all of the work that NBS had undertaken in the development of its standard, ANSI adopted exactly the same algorithm, known inside of ANSI as the Data Encryption Algorithm (DEA). Apparently the issue of key size would not seriously emerge again—judgment regarding that matter was being left to NBS, which had mustered as much expertise in open cryptography as any organization could.

Other ANSI committees, including the committee on Retail and Banking and the Financial Institution Wholesale Security Working Group—saw the adoption of DEA and established their own requirements to use the same Data Encryption Standard produced by the NBS effort.

In view of this activity, the American Bankers Association developed its own (voluntary) standard around the DES algorithm. The International Standards Organization (ISO) adopted the algorithm, calling it DEA-1. Australia's banking standard also was built around DES.

Given the widespread adoption of DES for data encryption, a great deal was at stake. If DES turned out to be resistant to serious attack, tremendous amounts of data being locked up in computers would be safe, even against the most sophisticated attacks. On the other hand, if anyone found an exploitable weakness or good attack against DES, tremendous loss would be possible.

Ruth M. Davis of NBS published an article in the November 1978 issue of *IEEE Communications Society* about the process of forming the Data Encryption Standard.[12] In it, she wrote that the workshops determined that DES was satisfactory as a cryptographic standard for the next ten to fifteen years. Interestingly, she specifically observed that, "the risks to data encrypted by the DES will come from sources other than brute-force attacks."

After DES was adopted as a standard, it would be subjected to many types of attacks, and its properties would be studied exhaustively. After years of cryptanalysis, consensus would emerge that DES was indeed a strong algorithm, remarkably resistant to a wide variety of attacks. Still, one criticism of the algorithm just could not be shaken. The key length, at fifty-six bits, was proclaimed insufficient to protect against attackers very far into the future.

Academic and industrial cryptologic research increased significantly in the years following the standardization of DES, including significant work done in the growing community of cryptographers outside of government intelligence agencies. Products would continue to be developed, with increasingly sophisticated systems becoming available and put into use. While not opening its vault of cryptologic secrets, the U.S. government did watch the ever-increasing size and sophistication of this community with great interest. The government's concern was not just with the independent domestic development of powerful new encryption products, but with the export of those products into the international markets.

As with other technologies that could raise national security concerns, the export of cryptographic products was subject to the International Tariffs in Arms Regulations ("ITAR"), administered by the Office of Defense Trade Controls at the Department of State. A license would be required for any U.S. companies or persons to export such items, and that license would be subject to approval of the State Department, which would presumably follow the recommendation of NSA.

The purpose of ITAR was to prevent the export not just of armaments but of implementations of cryptographic techniques. Working cryptosystems could only be exported outside of the U.S. and Canada with a key of forty bits or smaller, which would essentially mean that only systems that could be broken easily were allowed to be exported. There was no restriction on key length for domestic use, and by 1996 systems with keys of 128 bits and more were widely available. Even so, DES, which was already well-established as the de facto international benchmark, remained the standard for commercial usage.

4

Key Length

In any cryptosystem where a key allows the intended recipient to read the message, there is always a chance that an attacker will figure out which key will decrypt message. Longer keys are one of the simplest and most effective mechanisms to lower the risk: a machine that could find a fifty-six bit key every second would take 150 trillion years to find a 128 bit key. This is why Hellman and Diffie argued for longer keys; finding keys by trial and error would be simply ridiculous even to contemplate.

Cryptosystems are divided into two categories: symmetric (also called "secret key" or "conventional") and asymmetric (also called "public key"). In both categories, the concepts of key and the key length are of the greatest importance.

Symmetric cryptosystems use the same key for encryption and decryption. Physical locks are often symmetric.

A familiar example of a symmetric lock was mentioned on page 12: a bicycle chain with a combination lock that holds the two ends together until the numbers are rotated to display the proper combination. The key in this case is not a physical piece of metal, but the combination that the user can enter, which will cause the internal mechanisms of the lock to align so that the end pieces can be put together and pulled apart. If you know the combination, you can use the lock; if not, you can't.

All of these locks are vulnerable to an exhaustive key search, known as a *brute-force* attack. Attackers simply try every single possible combination until finding the one that works. Imagine a bicycle combination lock with one tumbler, with ten positions numbered from 0 to 9. The brute-force attack against this lock is to set the tumbler to position 0 and to pull on the lock to see if it opens. If not, move the tumbler to position 1 and pull on the lock to see if it opens. If not, systematically keep changing the tumbler position until you find the right one.

Such a system has a *work factor* of ten operations in the attacker's worst case scenario, meaning there is a one in ten chance of guessing correctly on the first try. On average, an attacker would be able to find the combination in five tries, assuming that the keys are distributed randomly.

One way to demonstrate random distribution, and the fact that on average we need to try only half of the keys to find the right one, is with a plain old six-sided die, the sides numbered 1 through 6. If we roll the die, each number has a one in six chance of coming up. Imagine that the die is being used to find the key for a tumbler with six positions, labeled 1 through 6, we'll be able to make the connection. Roll the die a large number of times—say, 100 times—recording which number comes up on each roll.

Now, if we set the one-tumbler, six-position lock to what comes up on the die, we have set the "key" for the system randomly, which is the best possible way to choose a key. If you then give the system to a group of attackers to unlock the system, they will probably set the lock to 1, pull it, moving on to 2 if it doesn't work, and so on, until they unlock it. The group can also try them all at random if they like. Even if the group employs both strategies, the result will be the same in the long run. If we record the number of attempts that it takes for the attackers to unlock it, we'll see that they have a one in six chance of guessing correctly on the first try. They have a six in six chance of guessing correctly through the sixth try. They have a three in six chance of guessing correctly through the third try.

If we assume that it takes one second to set the tumbler and to see whether the lock has disengaged, our ten-position, single-tumbler lock would be secure only against an attacker in a very big hurry.

If we want to increase the attackers' work factor, we can either increase the number of settings on the tumbler or we can add another tumbler. If we add another setting on the tumbler, we'll increase the attacker's worst case work factor to eleven seconds. If we add another ten-setting tumbler to the lock instead, we have increased the attacker's worst case work factor to 100 seconds.

Thus, increasing the number of tumblers is much more effective than increasing the number of settings on the tumbler. Figure 2 shows the possible settings on our lock with two tumblers numbered 0 through 9.

	x0	x1	x2	x3	x4	x5	x6	x7	x8	x9
0x	00	01	02	03	04	05	06	07	08	09
1x	10	11	12	13	14	15	16	17	18	19
2x	20	21	22	23	24	25	26	27	28	29
3x	30	31	32	33	34	35	36	37	38	39
4x	40	41	42	43	44	45	46	47	48	49
5x	50	51	52	53	54	55	56	57	58	59
6x	60	61	62	63	64	65	66	67	68	69
7x	70	71	72	73	74	75	76	77	78	79
8x	80	81	82	83	84	85	86	87	88	89
9x	90	91	92	93	94	95	96	97	98	99

Fig. 2. Possible Combinations For A Two-Tumbler Lock

Thus, by adding a new tumbler to the lock, the strength of the system is increased *exponentially*, whereas the strength of the system where a new position is added to the tumbler increases only *linearly*.

Mathematicians express this concept with simple notation like x^y, where x is the *base* and y is the *exponent*. A ten-tumbler system has a base of 10 and an exponent of the number of tumblers, in this case ten. Our first example, the single-tumbler lock has $10^1 = 10$ possible combinations. Our second example has $10^2 = 100$ possible combinations. A typical bicycle tumbler lock might have four tumblers, in which case there are $10^4 = 10,000$ possible combinations.

Still assuming that it takes one second to test each combination, it would take 10,000 seconds (nearly three hours!) to try every possibility on a four-tumbler lock. Once again, in practice, an attacker will only need to try approximately half of the keys on average to find the right one. So our system will be able to resist brute-force attacks for an average of just under an hour and a half.

Finding a cryptographic key is no different. In a brute-force attack against a cryptosystem, the attacker simply starts trying keys until one works. Since modern computers are binary, our cryptosystems are like tumbler locks with only two settings: `0` and `1`. Instead of saying how many "tumblers" we have in computer-based cryptosystems, we say how many "bits" we use to represent the key. A one-bit cryptosystem has two possible keys: `0` and `1`; mathematically, this is $2^1 = 2$. A two-bit cryptosystem has four possible keys: 00, 01, 10, and 11; mathematically, $2^2 = 4$. A three-bit cryptosystem has eight possible combinations ($2^3 = 8$).

What's interesting about breaking a cryptosystem, though, is that the equivalent of pulling on the lock to see if it opens involves running the encrypted message, trying a key that might unlock the message through the decryption process and then examining its output. The encrypted message will look like gibberish. Running the wrong key through the decryption process with the message will give us more gibberish. Running the right key through the decryption process will give us something that looks sensible, like English text.

For example, given ciphertext of `QmFzZTY0PyBQbGVhc2Uh` and the key `1101` as input, the decryption function would produce something like `UW1GelpUWTBQeUJRYkdWaGMyVWg` if the key is wrong. If the key is correct, the output would look like `ATTACK AT DAWN`.

This entire process can be automated with software. Consider a brute-force attack against messages encrypted with a three-bit cryptosystem. The software will need to recognize many popular data formats, for example, standard plain text, a JPEG graphics file, an MP3 sound file, and so on. To determine the key needed to unlock an encrypted message, the software would run the encrypted message through the decryption process with the first key, 000. If the output seems to match one of the known formats, the software will report 000 as a possible match. If not, it can go to the next key, 001 and repeat the process. Obviously, it won't take a computer long to work through eight possible combinations to find the right key.

The more strength we put into a system, the more it will cost us, so a balance must be struck between our own convenience—we can't make it too difficult for ourselves—and the attacker's. A lock that withstands attacks for an hour and a half is "secure" if it needs to protect something for an hour. A lock that withstands attacks for a year is "insecure" if it needs to protect something for a decade.

Team sports like American football provide a good illustration of the importance of timing in security matters. Football teams have playbooks, which are effectively code books. The quarterback calls the play, and his own team knows what to do next. The opposing team, on the other hand, should not be able to anticipate what the next play will be. If someone had the time before the play starts to analyze the quarterback's calls, their contexts (the down, how well the offense has been performing in its passing and running), and some history of the team's behavior, it's quite likely that he could figure out the play before it starts. But the code employed is secure because no one has time to perform all of that analysis. The message is secret for only a few moments, but it is enough time to serve its purpose.

There is one type of symmetric system that does not have the same weakness to brute-force attacks. This is the *Vernam Cipher*, developed by Gilbert Vernam of AT&T in 1918. Some—notably, Vernam himself was not one of them—have suggested that the Vernam Cipher is unbreakable, a claim which is worth considering.

The Vernam Cipher is actually a simple substitution cipher, one of the old manual systems (as opposed to modern computer-based systems) that used scratch paper and nothing else. Before we consider how the Vernam Cipher in particular works, we should be clear on simple substitution ciphers in general.

Julius Caesar is known for his use of a primitive encryption system that now bears his name. The Caesar Cipher is a simple mechanism of substituting one letter for another, following a regular pattern. To see how this works, write the alphabet:

```
A B C D E F G H I J K L M N O P Q R S T U V W X Y Z
```

Write the alphabet again, just below it, starting with N (shifting thirteen characters to the left).

```
N O P Q R S T U V W X Y Z A B C D E F G H I J K L M
```

The shifted-thirteen-places version of the alphabet is the key in the cipher. To encrypt a message with this system, simply choose the letter you want from the top alphabet, find the corresponding letter in the bottom alphabet, and write that down.

Thus,

```
ATTACK AT DAWN
```

becomes

```
NGGNPX NG QNJA.
```

Decryption works the same way; the intended recipient knows how to construct the bottom alphabet. When reading the message, he'll find the letter in message in the second alphabet and match it up to a letter in the first alphabet, revealing the original message.

Variations have been proposed, where instead of simply shifting the alphabet some number of spaces, the letters of a word like `QWERTY` are used to start the substitution alphabet. In such a case, the key would become

```
A B C D E F G H I J K L M N O P Q R S T U V W X Y Z

Q W E R T Y A B C D F G H I J K L M N O P S U V X Z
```

and

```
ATTACK AT DAWN
```

would become

```
QOOQEF QO RQUI.
```

A big problem with simple substitution ciphers is that they are vulnerable to a simple attack known as *frequency analysis*. The attacker counts how often each letter appears and compares that to how often each letter should appear in a message of a particular language. For example, in English, the most frequent letter is `E`, followed by `T`, so a cryptanalyst will begin by trying to replace the most frequently used letter in the encrypted message with the letter `E`.

That does not always work, as our sample message shows. In this case, the most frequent letter is `Q`, which stands for `A`. The second most frequent letter, however, does match the expected distribution. `O` is the second most common letter in the message, which corresponds to `T`, which is the second most common letter in English. Additional analysis will find other clues like letters appearing in double, certain letters that appear together, and the likely position of vowels and consonants.

The strength of a substitution cipher can be increased dramatically by replacing the substitution alphabet with a character stream invulnerable to frequency analysis.

The Vernam Cipher is precisely such a system. Rather than mapping one character to another using the same twenty-six characters, the Vernam Cipher relies on a key that is as long as the message that is being encrypted. So, if the message being encrypted is

`ATTACK AT DAWN`

the system will require fourteen random characters for the key. Those characters might be

`YDRJCN MX FHMU.`

Now the substitution can be made. To perform the substitution, we look to see how far into the alphabet each letter of the key is. For example, `Y` is the twenty-fifth character of the English alphabet, so we need to advance twenty-four positions into the alphabet to reach it. So we use the value twenty-four to perform the substitution.

Our plaintext for this digit is `A`, the first character in the English alphabet. Adding twenty-four to one gives us twenty-five, so the resulting ciphertext for that position is `Y`.

For the second digit, our key is `D`, the fourth character in the English alphabet, so we need to advance the next plaintext character three positions. Thus, our plaintext `T` becomes `W`.

In the third digit, our key is `R`, the eighteenth character in the alphabet, so we need to advance the next plaintext character seventeen positions. This `T` becomes `L`.

This process is continued until we arrive at the ciphertext of our message,

`YWLJEY MO IHIG.`

Decryption of the message requires the recipient have exactly the same one-time pad (the same key) and that the keys remain in perfect synchronization; if the sender starts to encrypt the next message with the thirteenth digit of the pad and the recipient starts to decrypt the message with the fourteenth digit of the pad, the result will be incomprehensible.

The correspondents in our example have been careful, so they are using the same pad and they're correctly synchronized. The recipient will take our ciphertext and the one-time pad, performing the same process as the sender, but in reverse.

Seeing that the letter `Y` in the ciphertext is zero digits off from the letter `Y` in the key is an interesting case. The recipient will look at the plaintext English alphabet, advancing zero digits before picking the letter. Zero digits, no advancement, gives the letter `A`.

Seeing that the difference between our ciphertext `W` (the twenty-third letter of the alphabet) and the key's `D` (the fourth character of the alphabet), we'll see there is a difference of nineteen. Reaching nineteen characters into the regular alphabet, we arrive at the letter T.

After examining the third character, we see that L in our ciphertext is the twelfth character of the alphabet. The corresponding digit in the key is R, the eighteenth character of the alphabet. Subtracting eighteen from twelve gives us negative six, which means we have to repeat the plain alphabet so we can move into that repeating series. (Of course, we can simply subtract six from twenty-six—the total number of characters in the English alphabet, giving us twenty—we will need to move nineteen positions into the alphabet to reach the twentieth character.) Advancing nineteen positions into our plain alphabet, we again arrive at T.

For the fourth character, we see that the difference between our ciphertext J and the key J is zero, so we advance zero digits into the plain alphabet, again yielding A.

Again, the process is repeated for each digit until we arrive back at the original message plaintext, ATTACK AT DAWN.

The safety of the Vernam Cipher depends on the key being perfectly random and used only once. Any reuse of the pad would open the way for attacks against the key (and therefore all messages ever encrypted with the pad), as was shown by NSA's VENONA project.

VENONA was a project started by the U.S. Army's Signals Intelligence Service—NSA's forerunner—in 1943 to monitor Soviet diplomatic communications. A brilliant success of American intelligence, VENONA gave the U.S. political leadership important insights into the intentions of the Soviet Union, as well as details of its espionage efforts.

VENONA is now best known for decrypting intercepted Soviet communications that showed Julius and Ethel Rosenberg's involvement in the Soviet spy ring. The project also showed how the Soviet Union gained information on U.S. atomic research, particularly the Manhattan Project.

Over the course of VENONA's lifetime, some 3000 intercepted messages encrypted by the Soviets with a Vernam Cipher were broken because the system was not used properly.[13]

Despite the restriction on the perfect randomness of the key and the difficulty in managing the key, the system really does have perfect security. Because the substitution happens against a randomly generated pad, a cryptanalyst cannot tell when the right sequence of decrypted characters is ATT or THE, because any digit has an equal probability of mapping to any other digit. Recall that frequency analysis requires a one-to-one mapping of a plaintext letter to a ciphertext letter, such that we can count how often a ciphertext letter appears and then to try reading the message by making the most frequent letter E, the next most frequent T, and so on.

Consider that when we attack a message using brute force in most systems, we can simply tell whether we found the right key because the result of the decryption process gives us something sensible, like plain English text, or something in a known data format like a JPEG image file.

The beauty of the Vernam Cipher is that when we find the key—which we will, if we perform an exhaustive key search of all possibilities—we'll have no way of knowing that it's the same key that was used by the users of the system. Consider the following example.

Given the ciphertext,

```
ASPSDFAQQPJF
```

all of the plaintexts listed below are equally plausible solutions, because they have the same number of digits, and each digit in the key has an equal probability of being any given letter, as a result of being selected randomly. (A mathematical property of a randomly-chosen character is that it has equal probability of being any particular element of the set. Returning to our example with the die, a fair die roll will mean that there is no difference in the odds of a 1 coming up and the odds of a 4 coming up.)

Consequently, a cryptanalyst who decrypts our ciphertext cannot tell whether our original plaintext message was any one of the following, or some other combination of letters equal in length:

```
CRYPTOGRAPHY

ATTACKATDAWN

BISCUITMAKER

THISISSECRET

HANDKERCHIEF

LEXICOGRAPHY.
```

Despite the tremendous security offered by the Vernam Cipher, the key length requirement prevents it from being used more widely. It is only manageable for communications that are brief and infrequent. (Rumor has it that a hotline established between Washington and Moscow after the Cuban Missile Crisis of 1962 was encrypted with a Vernam Cipher.)

With key management in symmetric cryptosystems, we have reason to be happy for the availability of asymmetric systems. In recent years, these have become popular, largely because the Internet has enabled many people who have never physically met or communicated through another means to exchange data, in the form of e-mail and through the Web. The basic idea behind asymmetric systems is that instead of one key, users have key pairs. One key in each pair is private, while the other is public, published as widely as possible.

If Alice wishes to send a message to Bob, she'll encrypt the message with Bob's public key. The only key that can decrypt a message encrypted with Bob's public key is Bob's private key. An interesting side effect of encrypting with Bob's public key is that Alice cannot decrypt a message that she herself wrote and encrypted.

Asymmetric systems tend to be more complicated, mathematically speaking, than symmetric systems. Symmetric systems are heavily dependent on jumbling messages through processes like transposition (where the data are rearranged) and substitution (where one datum is replaced for another). Asymmetric systems are generally based on mathematical functions that are easy to calculate one way, but extremely difficult to reverse.

A good example of a function easy to perform only one way is squaring. Taking any number, and squaring it is easy: simply multiply it by itself. Going backward—determining square root—isn't so easy. Consider $\sqrt{37,636}$, that is, what number, multiplied by itself is $37,636$? You'll need to take a guess and see if it works out to be too high or too low, then keep adjusting it until you find the right number. This takes a lot of time.

Compare the time spent in calculating square roots with what it takes, say, to compute 194^2. The difference in time for going forward (squaring) compared with going backward (taking square root) is what makes for a one-way function and demonstrates why an attacker has to spend so much more time and effort than users of a system who have the key.

By 1993, cryptography was becoming a more visible part of the emerging "Internet culture." As more people came online, concerns over privacy of electronic communication grew dramatically. Numerous attempts were made to find ways to secure Internet-based electronic mail. One group in particular was not about to wait for someone else to bring privacy to the world's electronic networks.

The second issue of a magazine called *WIRED* ran a black and white photograph of three masked men, holding up an American flag on its cover. The white masks bore numbers—one appearing as a bar code, the other two in hexadecimal: one across the top, and another across the eyes. Those numbers were the public key IDs of the faceless persons on the cover.

Over the photograph was the text, "Rebels with a Cause (Your Privacy)." Without explanation, the bottom-right corner carried the text "ПОДКЛЮЧАЙТЕСЬ!" (Russian for, "Be wired!")

In the corresponding article, the masked rebels were identified as the founders of a group known as *Cypherpunks.* Fiercely independent—taking marching orders from neither governments nor corporations—and strong believers in the power of cryptography, the Cypherpunks began to look at how to address the problem of protecting information from prying eyes.

Eric Hughes, a self-employed programmer in California and a co-founder of the Cypherpunks described the mission succinctly in his 1993 missive, "A Cypherpunk's Manifesto." He wrote,

> Cypherpunks write code. They know that someone has to write code to defend their privacy, and since it's their privacy they're going to write it... Cypherpunks will make the networks safe for privacy.

Cypherpunks was never a group with any kind of official membership. People simply called themselves cypherpunks and would show up at occasional physical meetings, while the bulk of their collaboration took place electronically—using the technology of the Internet to build systems designed to secure their communications. The mailing list became the virtual gathering place to discuss the politics and technology of cryptography, which is to say protecting liberty in the digital world.

August 1993
CRYPTO '93 Conference, Santa Barbara, California

Canadian cryptographer Michael Wiener shocked his colleagues with his presentation showing his designs for DES-cracking hardware. His scientific paper included detailed designs, as well as cost estimates for building various configurations of his DES cracker. A $1 million version of his machine would be able to break DES keys by brute force in three and a half hours. A $10 million version of his machine would be able to do the job in twenty minutes. A $100 million version would break DES keys in two minutes.

While Wiener had not actually built the machine, he did report that his chip was fully designed and could be fabricated for $10.50 each in quantity. Were someone willing to come up with the money, his machine could be built and deployed.

Inspired by Wiener's paper, Phil Zimmermann, the author of the free encryption system called PGP (Pretty Good Privacy) publicly speculated that large government intelligence agencies, particularly NSA, had budgets that would support $100 million computers. DES encryption was useless against the government.

Zimmermann knew that DES was not strong against a determined attacker years before Wiener's paper. PGP, developed over a period of some five years before its release in mid-1991 did not use DES. Instead, PGP used a 128-bit symmetric cipher known as IDEA—roughly seventy-two times more resistant to brute-force attacks than DES.

With Wiener's paper helping to quantify just how much money would be needed to build a computer that would render DES useless, Zimmermann wrote, "DES is dead, dead, dead."

October 1, 1996
Process Software Corporation, Framingham, Massachusetts

By late 1996, DES had been a standard for nineteen years. Several designs for machines to crack DES keys had been around for years. From time to time, someone would assert—without supporting any evidence, of course—that the government already had machines like the ones described early on by Hellman and Diffie and later by Wiener. Whether the rumors about government intelligence agencies actually having DES-cracking hardware at their disposal were true mattered less than the simple availability of designs that would crack DES-encrypted messages quickly and the fact that, like all computers, such machines were getting cheaper and easier to build.

Some attempts had been made to address the short key size in DES. In particular, several modes of operation were defined where a message would be encrypted more than once, with multiple keys. Two major standardized versions followed, both known informally as "Triple-DES" (or "3DES"), because a cryptographic operation was performed thrice. In the usual two-key variant, plaintext is encrypted with one key, run through the decryption operation with a second key, and encrypted again with the first key, giving an effective strength of 112 bits. In a three-key variant, plaintext is encrypted three times, each with a different key, giving an effective strength of 168 bits.

Despite cryptographers' lack of confidence in DES against determined attackers and the availability of stronger systems, DES continued to stand in official standards and was used very heavily. As long as the government stood by its standard, DES would remain in heavy use.

Cryptographer Peter Trei posted an article to Cypherpunks mailing list entitled, "Can we kill single DES?" "Single DES" was a reference to the original 56-bit standard for data encryption, to prevent readers from confusing it with Triple-DES. In his message, he determined that with the technology of the day, it was quite likely that DES keys could be broken at no cost in roughly six months. This would take into account the speed of computers in general use, the likelihood of how many participants would be willing to work on such a project, and just how much computing power would be needed to find one key out of 2^{56}.

5

Discovery

Summer 1985
Bexley Public Library, Bexley, Ohio

Like most cryptographers-to-be, I spent many of my summers off from school studying mathematical topics that I otherwise would not learn about until years later. I wanted to learn everything about how to manipulate numbers and symbols to determine the value of the mysterious x. Such studies necessitated frequent trips to the library.

At twelve years old, I discovered David Kahn's 1967 book, *The Codebreakers*, sitting on the shelf of the library I frequented. Being interested in history, mathematics, and computers, the book looked promising. With over 1000 pages documenting the role of cryptography in one decisive point in history after another, I found myself drawn into a world where secrets could be written and transmitted, incomprehensible to unintended readers. Just as intriguing were the unintended readers whose savvy and persistence allowed them to turn the jumble of ciphertext into valuable plaintext.

Before long, I became fascinated with the making and breaking of ciphers. Not content just to read about ciphers, I started to read more about how ciphers worked, and started to write my own enciphered messages—starting with the Caesar Cipher. Reckoning that a 2000-year-old cipher must be universally known, I began to develop my own systems for encrypting messages.

My first ciphers were intended to be manual systems—allowing for messages to be enciphered and deciphered with no more advanced technology than a pencil and paper. Having practiced my own skill at breaking messages enciphered with simple substitution systems (such as in the daily newspaper's "cryptoquote"), I started to look for ways to make substitution ciphers more resistant to attack.

In an effort to resist frequency analysis, I found that I could normalize frequency of letters and numbers appearing in ciphertext if instead of creating a simple one-to-one relationship between plaintext and ciphertext (e.g., where `A` is always written as `M`), a plaintext character could be mapped to a group (e.g., where `A` could be written as `M` or `1`). Letters with higher plaintext frequency would be assigned larger groups.

That led to a new problem: creating a mapping where some plaintext letters had more than one ciphertext equivalent meant that I needed more than twenty-six ciphertext characters to represent twenty-six plaintext letters. This problem was initially solved by adding numbers to the mix. By requiring that all plaintext messages have numbers spelled out, I could gain an extra ten ciphertext characters, thus allowing more ciphertext letters to represent common letters like `E` and `T`.

I kept a notebook of my ciphers, giving each a name, along with some comments on the system's strength and likely scenarios where each would be appropriate to choose. At the same time, I established a means to identify which cipher was in use.

Some of the more sophisticated systems were further resistant to analysis, using a single ciphertext digit for plaintext letters that commonly appear together, such as `TH`, `CH`, and `SH`. As I continued to develop these, I moved further away from simple substitution ciphers that change ciphertext to plaintext one letter at a time and into codes, where a symbol could mean a whole word or phrase. Such systems demanded more and more symbols to represent ciphertext, so I adopted the use of "codegroups" (groups of numbers or letters) for ciphertext. Once I had access to a computer, where I could create and manage such systems more easily, I settled on a system of five-digit numbers for my codegroups. Instead of being limited to thirty-six characters to represent ciphertext, I suddenly had 100,000. I could make the groups for common letters as large as I liked, add a large set of "nulls." Nulls were codegroups that had no plaintext equivalent: they existed only to confuse attempts analyze the ciphertext.

Encrypting a message with such a system was tedious because such a large table was needed. In fact, it would work very much like a foreign language dictionary that goes from one language to another in the front part, and then back again in the second half.

To encrypt a message like:

```
Meet at the library after school today. The topic is
algebra.
```

the plaintext-to-ciphertext table would be consulted to find words or phrases that would be defined. Most of the words in a message like this would be defined in their entirety, such as "meet," "at," "library," and "today." The phrases "after school" and "the topic is" would be defined in their entirety as well. Definite articles in this case can be safely dropped without any loss in meaning, leaving us to spell out only "algebra."

A few sample entries from the table below will give us what is needed to convert the plaintext to ciphertext.

```
.             34134
a             34714 10398 27225 36404
after school  72191
at            47008
b             88420
library       02979
e             16628 72700 33525 73487 69419 15940
g             30148
l             46399
meet          40746
null          22604 24638 84879 71792 03858 11341
              37135 11337 21715 92643 16989 67646
              82998 07445 83430 10869 15653 17431
r             66911
today         65573
topic is      08565
```

Working through our cleartext message, constructing the ciphertext from the codebook would create ciphertext that obfuscates not only the letters and words used to create the message, but its general structure. The resulting ciphertext would look like the following.

```
40746 47008 02979 72191 65573 34134 08565 22604
34714 46399 30148 16628 88420 66911 10398 24638
```

Decryption would use the corresponding table that was sorted by number, e.g.,

```
02979    library
08565    topic is
10398    a
16628    e
22604    null
24638    null
30148    g
34134    .
34714    a
40746    meet
46399    l
47008    at
65573    today
66911    r
72191    after school
88420    b
```

Using that table, the ciphertext could be converted back to plaintext, one codegroup at a time. After I began to use computers to construct these elaborate codebooks and to use them more efficiently, I learned how data are represented in computers and how they could be manipulated in much more powerful ways, opening a whole new world of cryptographic capabilities. It was then, in high school, that I learned about cryptosystems like DES and the mathematical methods used to try to break them.

I had not realized the complexity of the field until I learned enough that my pursuit of cryptography was no longer an absurdity.

My early cryptosystems—the most sophisticated of which I thought to be fiendishly devious—were all a joke in the grand scheme of things. On the other hand, they were secure for their purpose. My friends and I were able to use my systems to keep our secret plans to rendezvous at the library safe from the prying eyes of our "adversaries." But the experience was valuable and ultimately helped me to develop expertise in cryptography that would allow me to protect information from significantly more sophisticated cryptanalysts than teachers and parents.

6

RSA Crypto Challenges

August 12, 1982
MIT, Boston, Massachusetts

With signatures applied to a handful of documents, RSA Data Security, Inc. came to life, a company formed to bring the work of three MIT mathematicians—Ron Rivest, Adi Shamir, and Len Adleman—to market. The trio invented a public key cryptosystem in 1977 and named it "RSA" after themselves. Looking for a means to bring their algorithm into use, Rivest, Shamir, and Adleman formed the company and, with some financial backing from an experienced businessman, acquired the exclusive rights from MIT to commercialize their invention, U.S. Patent 4,405,829, "Cryptographic Communications System and Method."[14]

By 1996, RSA Data Security grew to be a company whose technology was used to protect all manner of information, especially on the Internet. Securing e-mail messages and Web transactions became everyday events thanks to products incorporating RSA Data Security's cryptographic algorithms. Such systems needed not only the flagship RSA public-key algorithm, but symmetric cryptosystems as well. To answer this need, RSA Data Security offered symmetric algorithms, including a system devised by MIT mathematician and RSA founder Ron Rivest.

The RSA algorithm itself depends on a difficult mathematical problem: factoring of large numbers (i.e., finding all of the smaller numbers that will divide evenly into a given large number). RSA key pairs are generated mathematically, and need to be made with big numbers—numbers so big that factoring them would take so long that by the time an attacker factored the number, the encrypted message contents would be worthless. Knowing how big those numbers had to be would require staying abreast of just how quickly large numbers could be factored.

As part of its ongoing research efforts, RSA Data Security sponsored a series of contests, known as the RSA Factoring Challenges. Those contests were designed to encourage researchers to factor large numbers—over 100 digits long—and to compete against each other to see who could factor the largest numbers in the least time. These efforts, in turn, helped RSA researchers keep an eye on how quickly the code-breaking technology was progressing, so that the code-implementors would be sure to use keys large enough to keep the real attackers at bay.

Scientists had factored large numbers on computers before. In 1988, an effort to factor a large number was undertaken by Arjen K. Lenstra and Mark S. Manasse. While large numbers had been factored before, the Lenstra-Manasse effort was unique in that it used the Internet to coordinate its participants' efforts. Rather than having the computations performed by one large computer or by many computers from the same lab, volunteers in the project ran special software on their computers that would have them work with all of the other computers running the Lenstra-Manasse software.

RSA Data Security Inc. issued its Factoring Challenge contests to ensure that the factoring work would continue to be done. The strategy worked. In 1993, RSA-129 (a 129-digit number) was successfully factored by 600 volunteers. Since that time, many other RSA numbers have been factored, including RSA-160. (RSA paid cash prizes to contest winners.)

In each of these challenges, the contestants would be given some encrypted data, along with some information like which cryptosystem was used for the encryption and a little bit of the text that would match properly-decrypted plaintext. In short, this provided enough information for researchers to be able to perform a "partial known-plaintext attack," a very realistic sort of attack, where a cryptanalyst knows a little bit of the decrypted message. As the cryptanalyst finds possible breaks, he checks his work against what known plaintext is available to see whether the break is real.

October 1996
Process Software Corporation, Framingham, Massachusetts

Cryptographer Peter Trei was known not only for development of cryptographic software, but for his public discussion of cryptography, its uses, and the impact that it could have on the lives of its users. Many read Trei's posts to the Cypherpunks mailing list with interest. "Can we kill single DES?" was a compelling question. Knowing that breaking a message protected by a cryptosystem as deeply entrenched as DES would cause a stir, Trei wisely argued that any efforts to defeat DES must be oriented toward a legitimate target.

In response to his message to Cypherpunks, Trei received private mail from "the R in RSA," Ron Rivest, suggesting that he contact Jim Bidzos, the president of RSA Data Security, Inc. Even respected cryptographers can sometimes be mistaken for kooks—not everyone knows where to draw the fine line between genius and insanity, so Trei hesitated before finally writing Bidzos. When he finally wrote Bidzos, Trei suggested some sort of sponsored contest, along the lines of the Factoring Challenges that RSA had been doing over the years.

Jim Bidzos responded to Peter Trei's suggestion quickly and enthusiastically. People from RSA Data Security began to work with Trei on constructing a legitimate attack—a sponsored public contest, following the model of RSA's Factoring Challenges—on the global standard for data encryption.

January 1997
Megasoft Online, Columbus, Ohio

In the eleven years since my discovery of *The Codebreakers*, I continued my study of cryptography and information security. After developing expertise in operating systems and network implementation, I worked on software security in the financial services industry. After working on computer and network security at AT&T Bell Laboratories, I landed at an early Internet start-up company called Megasoft Online. My job there involved security for our "Web Transporter" product, which managed software distribution and installation safely—all over the Internet, without the need to use floppy disks or CD-ROMs. That meant using cryptography and I was happy to put my experience to good use.

Like many professional and amateur cryptographers, I received an e-mail from Trei in January of 1997 in which he updated the cryptographic community on his progress in getting a DES message cracked. He told us about RSA's support and announced his DES Key Recovery program, DESKR.

DESKR was written for more recent releases of the Windows operating system, such as Windows NT and Windows 95. Most individual computer users would be able to run DESKR. Windows 3.11—the older version of Windows that ran atop of Microsoft's text-based DOS—was waning in popularity to the point that it seemed pointless to go through the extra effort to get the software to work on it. Cryptographer Steve Gibbons adapted DESKR to work on two Unix-based systems more frequently found in server systems in data centers, computers running the Ultrix operating system from Digital Equipment Corporation (DEC, which later merged with Compaq, which itself later merged with Hewlett-Packard) and from IBM.

Trei sent me a copy of DESKR, including Gibbons' adaptations. Since I wanted to run DESKR, I made additional modifications to the software to run on the systems that I had—a process known as "porting"—and send the changes back to Trei. My adaptations enabled DESKR to run on Linux, IRIX (the Unix variant from Silicon Graphics, Inc., known simply as SGI), Solaris 2 (the most recent Unix flavor from Sun Microsystems, Inc.), SunOS 4 (Sun's older Unix), and various BSD Unix platforms.

DESKR was ready for some ten different types of computers by the end of the month, when RSA Data Security launched its contest to crack a DES-encrypted message.

January 28, 1997, 9:00 A.M.
Sixth Annual RSA Data Security Conference, San Francisco

RSA president Jim Bidzos officially launched the "1997 Secret Key Challenge," a series of contests designed to test how quickly messages encrypted with various key lengths can be broken by brute force. RSA's popular annual conference was a perfect springboard from which to launch the contest. A press release was issued and some members of the media were there for the conference. Thirteen contests were announced, challenging participants to break messages encrypted with RSA's popular variable-strength cipher, RC5. The first contest was a $1000 prize for breaking a message protected by 40-bit RC5, a $5000 prize for the 48-bit RC5 contest, and a purse of $10,000 went to anyone for decrypting the message in the other contests which ranged from 56-bit to 128-bit configurations of the RC5 cipher.

RC5 wasn't the only target of the Secret Key Challenge. In addition to the twelve RC5 contests, a contest to crack a DES-encrypted message was included. Its prize was set at $10,000.

12:30 P.M.
University of California, Berkeley

UC Berkeley graduate student and cryptographer Ian Goldberg read the output from his program, designed to find a solution to RSA's 40-bit challenge. Running on the Network of Workstations (NOW) at UC Berkeley, Goldberg's program pooled together the unused processing power of about 250 workstations, testing approximately 28 million keys per second.

Goldberg grinned as he read the message on his screen.

> The unknown message is: This is why you should use a longer key.

Just three and a half hours after the launch of RSA's Secret Key Challenges, the 40-bit contest was over.

February 10, 6:52 P.M.
Swiss Federal Institute of Technology (ETH), Zurich

Germano Caronni, a graduate student working on a Ph.D. in communications and security, was distributing software for use by individuals on their own machines to try to find the solution to RSA's 48-bit challenge. Once started on a participant's machine, Caronni's software would make a connection to his keyserver—a computer that would tell the computer which keys to try. Once the computer got the message that said where to start and where to stop searching, it would begin working. If the computer did not find the key by the time it had tried the entire set (or "block") given to it by the keyserver, the computer would report back to the server that it had tried all of the keys in the range that it had been given and ask for another range.

Germano Caronni happily saw his system report to him:

> The unknown message is: The magic words are Security Dynamics and RSA.

Just over thirteen days after the start of the contest, Caronni's project found the winning key. Caronni felt a sense of vindication, finding the solution, after having been beat to the right key by Ian Goldberg on the 40-bit RC5 contest.

Caronni won $5000, which he donated to the non-profit Project Gutenberg, an organization creating, maintaining, and distributing electronic texts of books whose copyrights have expired.

I could feel a real sense of excitement building within the cryptographic community. Like hundreds of others, I ran Caronni's key-cracking software on a dozen or more computers to which I had access, trying to bring the project to a successful conclusion as quickly as possible. Though I had helped to make Peter Trei's DESKR software available to more computer types, I decided not to work on the DES Challenge until the easier 40-bit and 48-bit RC5 Challenges were answered.

Now it was obvious knew that these systems were weak and could be broken at no cost with a small investment of time. Once the 48-bit challenge had been answered, cryptanalysts returned their attention to the U.S. Government standard of nineteen years, DES.

By working to break a message encrypted with DES, cryptanalysts were doing much more than answering a contest call or engaging in a theoretical exercise. DES was *the* standard in virtually every industry in virtually every nation. It had been criticized from the beginning as being weak against a determined adversary because of its small key size. The time for theoretical designs, postulations, and estimates was over. It was time to show the world that it was possible to break a DES-encrypted message.

No one seriously believed that that attacking DES by brute force would be easy. Though finding the right key would just be a matter of time, the risk that presented itself was that if a cryptographic attack took too long to find the right key, we strong-cryptography advocates might undermine the very point we needed to make. If the project that found the key for the DES challenge proceeded at the speed of the 40-bit challenge, the search would take twenty-six years. If the project proceeded at the speed of the 48-bit challenge, the search would take nine years. Finding the right DES key could be the largest computation ever performed, and if we were going to succeed, we were going to need a lot of computing horsepower to share the burden.

Once the 48-bit challenge group finished celebrating its success, those of us who worked on the Caronni project moved quickly on to breaking DES, renaming the project DES-Challenge. We set up mailing lists. We discussed the architecture—the design for getting key-cracking software running on thousands of machines to coordinate—used by Caronni at length, its virtues, and how we might need to augment it in order to answer the challenge successfully in a reasonable amount of time. Graphic logos were made for people to put on Web sites to raise awareness and to recruit "clients" (individual processors running the DES key cracking software). We talked about how to build the DES cracking software and how to get all of the clients working with each other.

I wrote to Nicholas Petreley at the trade magazine *Infoworld* and described the work we were engaged in as well as why we believed that participating in the DES-cracking contest was so important. Based on the success of the 40-bit and 48-bit challenges, I estimated that DES keys could be broken in three to four months of effort by a dedicated attacker with no special equipment. My estimate, more optimistic than the one Peter Trei had proposed several months earlier, was based on the number of people who were interested in proving that small keys were inherently unsafe keys.

We weren't the only ones estimating how long it would take.

7

Congress Takes Note

March 20, 1997
Capitol Hill, Washington, DC

Howard Coble, a sixty-six year-old representative from North Carolina leaned forward to speak into his microphone. Coble, chairman of the House Judiciary Subcommittee on Courts and Intellectual Property, called the meeting to order.

The 105th Congress was debating several bills regarding cryptographic policy. The House was considering a bill called the "Security and Freedom Through Encryption (SAFE) Act," the primary purpose of which was to relax control over cryptographic technology used in the U.S. and by U.S. citizens abroad. A similar bill called Pro-CODE was working its way through the Senate.

Law enforcement officials widely thought that restrictions on cryptographic products and access to cryptographic keys were vital weapons against crime—both online and offline. Officials feared that cryptography that the government cannot break would cause them to lose their struggle against terrorists, drug traffickers, and child pornographers—the three most oft-cited criminal elements in these debates. Since the SAFE Act and Pro-CODE bills would largely eliminate governmental regulation over cryptography, the Clinton administration opposed these bills.

Government regulation of cryptography prevented U.S. companies from exporting their security products to customers in other countries. These vendors opposed continued regulation, arguing that they were unable to compete in a global marketplace against foreign companies that were not subject to similar restrictions. Additional opposition to cryptographic restrictions came from civil libertarians who worried that the eventuality of such regulation would be a police state in which the freedom of U.S. citizens would take second place to government interests. Computer scientists and engineers argued that the systems requiring government access to keys would actually reduce overall security, since failing to protect medical, financial, and personal information with cryptography would make criminals better able to steal such information.

This debate over the liberation of cryptography started in the early 1990s and came to be known simply as the Crypto Wars.

Laying groundwork for the discussion to follow, Coble said,

> Today the subcommittee is conducting a hearing on H.R. 695, the Security and Freedom Through Encryption (SAFE) Act, commonly known as the SAFE Act. H.R. 695 addresses the complex and important issue of encryption.
>
> Encryption, as you perhaps know, is the process of encoding data or communications in a form that only the intended recipient can understand. Once the exclusive domain of the national security agencies, encryption has become increasingly important to persons and companies in the private sector concerned with the security of the information they transmit.
>
> The encryption debate encompasses two main issues. The first is whether there should be any restriction on the domestic use and sale of encryption technology and, in particular, whether domestic users may place their keys in escrow with the government or some neutral third party. This requirement would provide a mechanism which would allow law enforcement and national security agencies some ability to monitor transmissions. Current law does not have such restrictions.
>
> The second issue is whether there should be any restrictions on the export of encryption technology. Current law regulates the export of encryption technology in a manner similar to military technology.

After hearing from two representatives, the committee heard testimony from William A. Reinsch, undersecretary at the Bureau of Export Administration in the U.S. Department of Commerce. Reinsch described the Clinton administration's policy on cryptography: availability of strong cryptography to protect commercial and personal interests without sacrificing the ability to investigate alleged crimes and to protect national security.

Implementation of the Clinton policy for encryption was centered around key recovery and key escrow systems. Such systems make it possible for an authorized agent to decrypt ciphertext to reveal the original plaintext. While this is what decryption keys normally do, key recovery systems are designed to allow decryption even if the key is lost or if the key holder refuses to divulge it. Key escrow systems have a similar end result, but their mechanism relies on a trusted third party, one who essentially holds a "master key" that can be applied to decrypt ciphertext.

As part of the administration's attempts to implement this policy, cryptographic products were recategorized. Instead of being considered a munition and thus regulated by the Department of State, cryptography would be recognized as a dual-use technology (one that would be use in normal commercial activity in addition to government or military operation), and thus regulated by the Department of Commerce.

In addition, export restrictions were altered, temporarily changing the export limit from forty bits to fifty-six bits. This change explicitly allowed DES and "equivalent products" to be exported, provided that the exporting company submit plans to show they were working to develop a "key management infrastructure" (essentially, key recovery or key escrow). After a two-year transition period, exporters of cryptographic technology would be expected to have their systems support the sort of key management infrastructure envisioned by the administration.

Despite the relaxation of cryptographic product control, the administration did not want simply to let go of cryptography. Reinsch made the administration's view quite clear in the conclusion of his testimony.

> I must tell you that legislation such as H.R. 695 would not be helpful, and the administration cannot support it. The bill has a number of similarities to what we will shortly submit, but it proposes export liberalization far beyond what the administration can entertain and which would be contrary to our international export control obligations. We are sympathetic to some aspects of H.R. 695, such as penalties for unlawful use of encryption and access to encrypted information for law enforcement purposes, but the bill does not provide the balanced approach we

> are seeking and as a result would unnecessarily sacrifice our law enforcement and national security needs. I defer to other witnesses to describe the impact of the bill on law enforcement, but let me describe a few of its other problems.
>
> The bill appears to decontrol even the strongest encryption products, thus severely limiting government review of highly sensitive transactions. The administration has a long-standing policy that the risks to national security and law enforcement which could arise from widespread decontrol of encryption justify continued restrictions on exports.
>
> In addition, whether intended or not, we believe the bill as drafted would preclude the development of key recovery even as an option. The administration has repeatedly stated that it does not support mandatory key recovery, but we most certainly endorse and encourage development of voluntary key recovery systems, and we see a strong and growing demand for them that we do not want to cut off.
>
> As I have said on many occasions, Mr. Chairman, encryption is one of the most difficult issues in public policy today, but we are committed to solving it in cooperation with industry, the law enforcement community, and the Congress in a way that reinforces market principles and achieves our diverse goals. We hope that you will work with us to facilitate that process by passing the legislation we are proposing.

Next to testify was Robert S. Litt, a deputy assistant attorney general in the Criminal Division of the Department of Justice. After weighing in with the Department of Justice's view on the proposed legislation, Litt also offered his thoughts on the strength of cryptographic keys. Litt's remarks were focused on the systems already in use, rather than the sorts of key escrow systems that other administration officials were discussing.

Litt began with a high-level description of the argument put forth by some citizens. That argument held that the fears of U.S. law enforcement and intelligence agencies were "overstated," and that the government simply did not want its own citizens communicating in a way that would keep them safe from governmental eavesdropping. In effect, the government would oppose any cryptography that it could not easily break. (Cryptosystems for general export were limited to 40-bit keys. As William Reinsch had pointed out earlier, companies willing to show how they were going to implement key recovery or key escrow systems would be allowed to export cryptosystems with up to 56-bit keys.)

As evidence for the argument that only weak cryptography would be allowed, many advocates pointed to Ian Goldberg's victory in RSA's 40-bit key cracking contest. If Goldberg could break messages protected with 40-bit cryptography in three and a half hours, the argument went, the government must have the ability to break those messages as if they were not encrypted at all.

"This argument does not withstand scrutiny," said Litt. Pointing out that the computational power needed to decrypt a message by brute force rises exponentially as the key size increases, Litt attempted to show how brute force attacks simply were not an option for law enforcement.

"According to the National Security Agency's estimates, the average time needed to decrypt a single message by means of a brute force cryptoanalytic attack on 56-bit DES—a strength whose export we are now allowing—would be approximately one year and eighty-seven days using a $30 million supercomputer."

The law enforcement message to the U.S. Congress was unambiguous: brute force attacks against DES were infeasible.

The day's estimates would not stop there. William P. Crowell, deputy director of National Security Agency was next to testify. Crowell began, "I appreciate the opportunity to comment on the pending ... legislation and to discuss with you NSA's involvement with the development of the administration's encryption policy. Since NSA has both an information security and a foreign signals intelligence mission, encryption touches us directly." He went on to describe how NSA was acting as a technical advisor on cryptography to the administration.

In his testimony, Crowell said that the use of cryptography can be of significant benefit to the nation. From there, he outlined key management infrastructures and how public-key cryptography works, the need for an infrastructure to support public key cryptography, and how such infrastructures can support key recovery.

Finally focusing on the most vocal part of the cryptography debate, Crowell said, "I would like to help clarify some of the frequently-repeated factual errors regarding encryption so we all can stand on firm ground during the formation of the nation's encryption policies."

Crowell argued that basing long-term cryptographic policy on key size and brute-force attacks is shortsighted. Addressing this matter directly, he said,

> You may have heard news accounts of a University of California, Berkeley student who recently decrypted a message that was encrypted with a 40-bit key using 250 workstations as part of a contest from RSA Inc. This so-called "challenge" is often cited as evidence that the government needs only to conduct "brute force" attacks on messages when they are doing a criminal investigation. In reality, law enforcement does not have the luxury to rely on headline-making brute force attacks on encrypted criminal communications. I think you will find it useful to see for yourselves how increased key sizes can make encryption virtually unbreakable. Ironically, the RSA challenge proves this point.
>
> If that Berkeley student was faced with an RSA-supplied task of brute forcing a single PGP based (128-bit key) encrypted message with 250 workstations, it would take him an estimated 9 trillion times the age of the universe to decrypt a single message. Of course, if the Berkeley student didn't already know the contents of part of the message RSA provided some of the unencrypted message content to assist those who accepted the challenge it would take even longer.[15]
>
> For that matter, even if every one of the 29,634 students enrolled at UC Berkeley in 1997 each had 250 workstations at their disposal 7,408,500 computers (cost: $15 billion) it would still take an estimated 100 billion times the age of the universe, that is over 1 sextillion years (1 followed by 21 zeros), to break a single message.
>
> If all the personal computers in the world, 260 million computers were put to work on a single PGP-encrypted message, it would still take an estimated 12 million times the age of the universe, on average, to break a single message (assuming that each of those workstations had processing power similar to each of the Berkeley student's workstations).
>
> Clearly, encryption technology can be made intractable against sheer compute power, and long-term policies cannot be based on bit lengths. Brute force attacks cannot be the primary solution for law enforcement decryption needs. This line of argument is

> a distraction from the real issues at hand, and I encourage you to help put this debate behind us.

Crowell's argument was an interesting one. While he intended it to be taken as evidence that brute force attacks against commonly-available cryptosystems were simply not feasible, those who argued for freedom in cryptography would interpret Crowell's words much differently. To them, NSA's position suggested that the government was reluctant to allow its citizens to engage in free speech and virtual association via global networks without the prying eyes of even the most powerful government agencies.

Crowell's argument was also interesting from a technical point of view. While he used a recently-publicized event to provide estimates on how long it would take to crack a key by brute force, he used the speed of Goldberg's 40-bit challenge solution, rather than Germano Caronni's 48-bit challenge solution—even though Caronni's was considerably faster.

Another critical element of Crowell's argument was that it assumed that available computing power would remain constant—ignoring Moore's Law, which essentially says that computing power doubles every eighteen months. Thus, a computation that might take two months in early 1997 with "current technology" would take one month in mid 1998, and be down to two weeks at the beginning of 2000.

To many private cryptographers, it would appear that, just as it had twenty-five years earlier, the government was overstating the difficulty of brute force attacks.

8

Supercomputer

Testimony before the House Judiciary Subcommittee on Courts and Intellectual Property was compelling across the board. Everyone seemed to agree that the stakes were high and that breaking encrypted messages by brute force was a hard, time-consuming problem, even for well-funded government agencies.

The Justice Department's Robert S. Litt provided some of the day's most interesting testimony, not only providing estimates on the difficulty of cracking a cryptosystem by brute force, but specifically providing an estimate for cracking DES keys. Litt even cited the source of his estimate—NSA, the very same intelligence agency responsible for the brilliant cryptanalysis that uncovered Soviet spies operating in the United States after World War II. If anyone understood cryptography, it would be NSA.

NSA's estimate, he said, was that even with a $30 million supercomputer it would take a year and several months to decrypt a DES-encrypted message by brute force. Litt's argument was especially strong, drawing on the common knowledge that supercomputers were the fastest and most powerful computers available. Indeed, supercomputers were very good at dealing with very complex problems, tracking huge amounts of data, and working with gargantuan numbers.

But finding cryptographic keys in a brute-force attack isn't a large, complex problem. A search wouldn't need many data and the numbers involved weren't very big, at least as far as computers were concerned. Finding a key by trying every single one until the right key is discovered was really a large number of very small problems. The security of the system relies in the sheer number of keys that must be tested to find the one that unlocks the message.

You might think about a test that would require you to solve arithmetic problems like $1 + 3$ and $9 + 4$. Those problems aren't difficult at all, but if you must finish the test in an hour and there are one million problems, you might not be up to the challenge.

There are several ways that you can increase your chances of success. You could have all of your friends work on the test with you, giving each person a separate sheet of paper with some of the arithmetic problems on it. Perhaps your friends might recruit their friends, and you could increase the number of people helping you on your test further still. Having a mathematician join the project isn't going to help you much, though. While a mathematician can perform much more complicated operations and can work with much larger numbers, a fifth grade student could solve simple arithmetic just as quickly as the greatest mathematician in the world.

Using supercomputers to find DES keys would be just as expensive and inefficient as using mathematicians to solve a large number of trivial arithmetic problems. Just as an army of fifth graders would be cheaper and more effective in finishing a million-problem test of arithmetic, a large number of regular computers would be much more effective than a single supercomputer in finding a cryptographic key.

The fundamental issue here is how easily the problem can be "parallelized"—broken into steps that can be performed simultaneously by different computers, instead of all in order, one at a time. Engineers often illustrate this problem by pointing out that bringing a new person into the world is not something that can be parallelized. One woman is pregnant for nine months before we have a new person. We cannot expect a baby at the end of a month by impregnating nine women.

Testing cryptographic keys can be easily parallelized, just like a large number of trivial arithmetic problems. To show how supercomputers and desktop computers compare in this contest, let's assume that a supercomputer can test a cryptographic key in one second and that a regular desktop computer can test a cryptographic key in ten seconds. One supercomputer (at, say, $30 million) will be ten times the speed of a single desktop computer (at, say, $3000). That same supercomputer would be the same speed as ten desktop computers ($30,000). The same supercomputer would be one tenth the speed of 100 desktop computers ($300,000).

Almost since the beginning of digital computing, our machines have become smaller, less expensive, and faster. Where a single, monolithic machine used to serve a user community, subsequent generations of computers have become more efficient and more numerous. By 1997, personally-owned computers were commonplace in homes and dorms, and those machines were many times faster than the kinds of minicomputers and even mainframes that were used a decade or more earlier.

At the same time, progress in telecommunications allowed computers to communicate with each other as they had not been able years earlier. The rise of the Internet enabled this trend to the point where every computer in the world had the ability to communicate with nearly every other computer in the world.

Part of what made the Internet special was that it was not a network of computers, but of *networks*. Having Alice put her machine on a network would allow it to communicate with others on that network. Bob putting his machine on a network allowed his computer to talk to others on that network. The Internet made it possible for Alice's and Bob's networks to connect to each other.

Though many academic and business environments had dedicated connections to the Internet, individual users would typically use modems to call an Internet Service Provider (ISP). The connection was only temporary, though, and would only stay active as long as the user needed to be able to exchange email, to surf the web, or to read articles.

Very large numbers of computers were out there, able to communicate with other machines on the Internet. The question was how to get them to cooperate with one another and apply their energies to a singular problem like finding the particular DES key needed to read the encrypted message, thereby solving the RSA challenge.

RSA didn't put limitations on how anyone could solve its challenges. With 40-bit and 48-bit RC5 challenges already solved, the world's cryptographers were thinking seriously about how to increase the efficiency of a large-scale key search computation.

Increasing the number of computers looking for the key is one way to increase the efficiency of the search, but two other important options are available to the cryptanalyst who wants to use a brute-force attack. The first is to increase the efficiency of the software itself, and the other is to wire the instructions needed to find DES keys directly into a computer's hardware.

Consider the first option—efficiency in software itself. This alone can have a dramatic improvement in performance. When people write programs for computers, they typically use languages that were created for the specific purpose of communicating an algorithm or program specification to a computer. Computers don't run the programs that are created by the human programmers. Instead, programmers will write software in a computer language like Java, Lisp, or C++. The version of it that programmers work with is called *source code.* When the software is finished, the programmer will invoke a program called a *compiler* to read the source code and turn it into a form that the machine can read and run directly, known as the *binary executable* or *object code.*

In essence, the compiler will translate what the programmer says into something that the computer can do. Building software this way is economical and frankly much more practical than having people write object code directly. By making the process of creating the software easier, we reduce the amount of "people time" needed to create a fully operational system and to get it running. Since general-purpose computers like your typical PC, Macintosh, or Unix system are so fast and cheap and people are so expensive, we ultimately save huge amounts of money having computers do things for us.

If a program was generated by a compiler, its speed is nowhere near what it would be if the programmer with the skill took the extra time to write the program in a language that the computer can understand.

A computer language known as Assembler makes it possible for a programmer that knows the machine very well to be able to create extremely efficient code which is tailored to a given machine and contains very specific instructions.

An intimate understanding of the computer will allow the programmer to write code that performs the job in very few instructions, taking advantage of any little feature that a particular computer has available.

Languages (like Lisp and Java, mentioned earlier) are typically called "high level" when they allow the programmer to work on a solution without thinking much about how the computer is actually going to get the job done. "Low level" languages (like Assembler) require the programmer to specify exactly how the machine will store and manipulate the data in order to solve a problem. The difference between high-level and low-level languages is roughly the same as saying "walk down the street with a little spring in your step" and saying precisely how much to move each muscle in the body in order to move down the street. Writing programs in Assembler is comparatively hard work. Relatively few programmers can write in Assembler, because it requires much more intimate knowledge of computer hardware—a level of knowledge that many programmers simply lack.

A program written so "close to the metal" can be extremely fast on the system it was designed for. Any variation in that system, though, can slow it down dramatically. Even an upgrade in processor type (from one Pentium model to another, for example) will alter the processor's internal structure, making a program that ran very quickly in the old system perform very badly in a newer and faster system.

Peter Trei was able to implement his DESKR program in relatively little time because he didn't take the time to make the software search for keys as fast as the machine could possibly do it. Instead, he chose the language that most cryptography software is written in, known as C, a sort of mid-level language that doesn't let the programmer completely forget about the hardware, but doesn't require him to specify exactly what to do with every single bit in the entire system. That flexibility also allowed Steve Gibbons to make relatively minor changes to the software so that it would work on his computers and for me to make other minor changes so it would work on my computers. If Trei had built his software in Assembler, we would have had to write some parts of the DESKR software almost from scratch to work on other computer systems.

The only question remaining was whether key-searching software written in C was fast enough to get the job done. If not, the new key-cracking system would have to be written in Assembler, or possibly something even faster.

Building a physical system to perform a brute-force attack in hardware would increase its efficiency further, making it even faster than Assembler. However, the issues that need to be addressed when implementing a program in Assembler are still present when implementing that same program directly in hardware. Even worse, hardware implementations require building or physically wiring special-purpose equipment, which is much more expensive than reprogramming software.

Whether to crack keys with regular software written in a high-level language, specialized software written in a low-level language, or with custom-built hardware comes down to how fast the system needs to find keys and how much time and money can be thrown at building the system.

Litt said a supercomputer would take over a year to crack a DES key. If it took too long to solve RSA's DES Challenge, the government would have proved its point about the difficulty of breaking encrypted messages by brute force and we would have no choice but to shut up. The public policy debate over the freedom to use strong cryptography in the United States could be heavily influenced by how cryptographers would respond to the DES Challenge. We had a lot of decisions to make, and we had to choose well.

9

Organizing DESCHALL

February 1997
Megasoft Online, Columbus, Ohio

Mindful of the computing power we would need to find DES keys, it was important to take advantage of the momentum that had been built up by the publicity of the 40-bit and 48-bit RSA contests. It made perfect sense for the same group of people who ran Germano Caronni's 48-bit key-cracking software and communicated on mailing lists to form a new group that would try to solve RSA's DES Challenge. These hobbyists, students, amateurs, and professionals from all over the world (though heavily European) were already "assembled" on the mailing list, already interested in the topic, and already prepared to take on RSA's next challenge.

Just like Caronni's 48-bit contest effort, there would be no formal organization, no headquarters, and no employees. People who saw work that needed to be done would simply do it. After the software was built, volunteers would run the key-searching programs on their own computers and communicate via e-mail—participating from wherever they happened to be, putting in as much time as they wanted. Because a number of participants were from European countries that lacked U.S.-style rules against the export of cryptographic software, many of us thought that the group would produce software that everyone could run. Even in countries where the *export* of cryptographic software was forbidden, there were usually no prohibitions on the *import* of cryptographic software from other countries. So if the software were built in Switzerland, for example, volunteers from the U.S. could run the software. If U.S. programmers wrote the software, sending it to European volunteers would probably be an illegal export of regulated technology.

Many of us who ran Caronni's clients wanted to move immediately to breaking DES keys. Although I had previously worked on Peter Trei's DESKR system, I wanted to help with a *coordinated* effort—one that would harness the computing power of many machines in an organized way to search for the right key.

Some participants in Caronni's effort volunteered to perform certain vital functions for the DES Challenge early. Six volunteers agreed to work together to direct the effort and to coordinate the activity of those who were interested in answering the DES Challenge. Germano Caronni agreed to be part of this group, as did Piete Brooks of Cambridge University, Jered Floyd of MIT, Tim Newsome of CMU and a student programmer at Megasoft Online, Thomas Roessler from Germany, and "Thomas S." from the University of Manchester Institute of Science and Technology.

As the "DES-Challenge Organisation Committee" announced itself and the work that was to be done, it unveiled a half-dozen mailing lists to foster communication among participants, each dedicated to some specific part of the project. On those new lists, cryptographers and enthusiasts from all over the world discussed what would be needed to defeat DES. The group quickly agreed that we would need a system that could support many, many more clients than we ever had working on RSA's 48-bit Challenge—which peaked at roughly 4500 at once (and roughly 7500 total working for any length of time) for a peak key testing rate of 440 million keys per second.[16]

After much discussion, a general project architecture emerged. Instead of depending on a single server—a single place where a failure could bring the whole effort to a halt—there would be hierarchies of servers, each dedicated to a specific purpose. Another server would handle just the highest-level coordination sitting on the network at the root of the server hierarchies. While someone—we didn't determine who—would be responsible for the root server, others would be able to run servers from wherever they were.

There would be separate servers to hand out key ranges to clients, while others would receive the status reports from clients. Additional servers would collect the statistical information from clients (e.g. what kinds of operating systems were being used, how fast different systems were running the key testing software, and how many keys had been tested). Still other servers would subdivide key ranges into smaller ranges for use by very slow clients. Finally, there would be servers that would convert messages between the key testing client software and the keyserver into a form that would traverse firewalls—devices designed to prevent unauthorized intrusion into networks; some options we discussed included sending the messages through e-mail and the Web.

Still, when it seemed that everything was set and the European DES-Challenge group effort was well on its way, traffic on the lists dried up. Before long, it became clear that the actual software needed to crack DES keys was not being written.

While the DES-Challenge team was busy debating technical details other DES key search projects, such as the DES Violation Group (based in the U.S.) and SolNET (based in Sweden), had gotten underway. As the DES-Challenge team endlessly debated the details of its complex scheme for allocating key space for the search, I received private e-mail from Justin Dolske, a graduate student in the Computer and Information Science Department at the Ohio State University. Though Dolske and I lived less than fifteen minutes away from each other, it was not until we both participated in Caronni's Internet-based project in Switzerland that we got to know each other. Dolske told me about a project being led by a programmer named Rocke Verser. Verser's project had no name, no Web site, no mailing lists, and no support staff. It only had key-cracking software. Fast software.

Looking at how the DES-Challenge team operated, it was now clear that software didn't spring from a project with a name and a mailing list. A fully-functioning project, however, could well spring from working software.

January 1997
Loveland, Colorado

Fig. 3. Rocke Verser, 1997

Like many freelance programmers, Rocke Verser had spent years honing his software skills, taking on a variety of challenges sure to make him better able to practice his craft. Long attracted to mathematics and the computational machines that allowed complex calculations to be made so quickly, Verser was naturally drawn to cryptography. Finding work in the development of cryptographic software meant that Verser understood DES and knew ways to make it work well.

Intimately familiar with Intel's processors, Verser had written programs to encrypt and to decrypt data using the DES algorithm in the kind of hand-crafted Assembler that very few people can understand, much less write. To make his programs run even faster, he would manually optimize his programs to run in ways that were counterintuitive but ultimately faster. When a program would need to perform a series of calculations, he would look for opportunities to reduce the number of steps needed by changing the order of the calculations in a way that would still produce the correct result. An example might be a program needed to calculate how many matches a league of soccer teams would need to hold to determine a winner. Many programmers would write software to list the teams and put them in pairs, creating a "tree" to represent the winner's bracket in memory and then having the software count the number of branches in the tree. Verser would give the problem some thought and determine that the answer was always the number of teams minus one and write software to find that solution. Finding shortcuts like this would result in a program that provides the correct result, but in a dramatically faster way than those written by others.

It had been some time since Verser had worked on his fast DES implementation when he saw RSA's announcement of the challenges. He pulled up his fast DES software, having found a way to put them to good use again.

After some additional development, Verser looked at Germano Caronni's 48-bit RC5 software. Caronni's system of messages that the clients and server would send back to each other—"protocol" in network parlance—was easy to implement, and it got the job done to solve the 48-bit RC5 Challenge.

Verser then wrote a keyserver—a software system to coordinate clients, telling them which section of the DES key space to search. He built the keyserver to use the protocol adopted from Caronni, made up of simple messages like, "Ready for keys," "Test block number...," "Finished block number...," and "Stop working."

Verser's previously-existing software and his decision to use a simple architecture allowed him to get a basic distributed DES key-cracking system up and running in relatively little time. Focused on building the system for testing DES keys, he didn't get distracted in trying to find an army of willing participants, setting up Web sites and mailing lists, or even finding a clever name for his project. It was all simply about RSA Data Security's DES Challenge. His clients were given such exotic names as `DESCHAL4.EXE` for Intel 486 processors, `DESCHAL5.EXE` and Intel Pentium ("586") processors. Verser's DES Challenge project would come to be known as DESCHALL.

By using the keyserver in conjunction with the DESCHALL key-cracking clients, Verser envisioned many of these clients interacting with the keyserver much like parts of a very large computer, spread out all over the United States. The keyserver would be the central coordinating unit, breaking all possible keys into blocks that could be handed off to the clients for testing. A person with Verser's client software on their computer would simply start the DESCHALL client, which would then ask the keyserver for a block of keys to test. When the client was finished testing all of the keys in that block, it would tell the keyserver it finished and request a new block to test. By communicating with the participants in this way, the keyserver would keep track of which keys had been tested and which remained to be tested. If the right key was found, the client would immediately and automatically contact the server to report the find.

Because of the efficiency of the protocol Verser implemented, the keyserver could be a modest machine. Verser had such a machine, with a 66 MHz Intel 80486 processor and IBM's OS/2 operating system. He implemented his keyserver in a high-level language known as REXX, originally written for mainframe computers and later adapted for machines like the Amiga and PCs running OS/2.

Thus, Verser had the opposite problem that the European DES-Challenge group had. Verser had a superb software system for cracking DES, but he didn't have anyone to help him find the right key.

Tuesday, January 21, 4:39 P.M.
Longmont, Colorado

Michael Paul Johnson, a programmer working independently on a variety of projects, had an interest in cryptography and had developed several cryptographic utilities that he wanted to distribute online.

Not wanting to break the law on cryptographic software export, Johnson developed a system that would allow his software to be distributed after providing notice that the software is subject to export restrictions and getting some verification that the user is allowed to download the software. Johnson's system came to be known as the "North American Cryptography Archive."

To comply with the Export Administration Regulation (EAR) that was in effect for cryptographic software, Johnson required users to enter a name and a password. If those checked out, users would be allowed to proceed to the archive itself, which was in a hidden location that changed every hour.

Users could get their own names and passwords for the archive by filling out an online form. Users would supply their e-mail addresses, and then answer three questions:

1. Are you and the computer(s) you are operating both in the United States of America or Canada?
2. Are you a citizen of the United States of America, a permanent resident (with "green card") of the United States of America, or a citizen of Canada?
3. Are you aware of the U.S. Export Administration Regulations and similar Canadian regulations?

Users who provided affirmative answers to all three questions and an e-mail address that appeared to be based in the U.S. or Canada received a name and password for the site in e-mail. Non-answers and e-mail addresses that did not appear to be from the U.S. or Canada would result in a denial of access to the archive.

Johnson described this system on a mailing list called Cryptography and included a generous offer to host cryptographic software for distribution to North America.

Rocke Verser took advantage of Johnson's offer and decided to use the North American Cryptography Archive as the distribution point for DESCHALL key-cracking client software.

Saturday, February 22
Megasoft Online, Columbus, Ohio

After talking with Justin Dolske about Rocke Verser's fast key-cracking software, I decided to contact Verser myself. After exchanging some pleasantries, Verser let me know where he had put the DESCHALL clients for download. I started running his key-cracking client software on about a half-dozen machines at my home office and a few more back up at our company headquarters in Freehold, New Jersey.

Word was quietly spreading among people interested in RSA's DES Challenge that Rocke Verser's software was really fast. By March, more people were running Verser's DESCHALL key-cracking clients. Verser, Dolske, and I traded e-mail daily, discussing how to get more people involved.

As was true with the other DES Challenge projects, the earliest DESCHALL participants were savvy about cryptography and computers and so they needed very little help to start participating. They handled minor differences in the way that their computers were connected to the Internet, sometimes after asking Verser for a little help. Nevertheless, most people who didn't know about cryptography had no idea that software like DESCHALL existed, and they had no idea why running a DESCHALL client to help someone win the RSA contest was a good idea.

Coordinating the efforts of a slowly growing number of people became increasingly difficult. Verser, Dolske, and I all agreed that we should have a public mailing list, where everything that we would write about the project would be sent, allowing participants to follow the conversation and to participate in it. (Of course, if we needed to communicate privately, direct e-mail was still an option.) By having a public mailing list, we could have people write to the mailing list when they needed help with the DESCHALL client software, allowing people other than Verser to answer such queries. This, in turn, allowed Verser more time to make improvements in the DESCHALL software itself.

Once a core group of participants had been established, we looked at the resources that each had available. The main Megasoft offices in Freehold had good Internet connectivity, and I ran an important machine there, named gatekeeper. Because gatekeeper was always connected to the Internet, it was the best place to host a public mailing list to help us coordinate our attack on DES.

Saturday, March 29
Megasoft Online, Freehold, New Jersey

As the sun set, I logged into gatekeeper in Freehold from my home office in Columbus. I installed a software package known as Majordomo, which would automatically run mailing lists and keep archives.[17]

Messages started to flow into the mailing list immediately. In the first hour and a half that the list was online, participants from all over the country started to enumerate the obstacles before us so we could set about finding solutions.

Ironically the biggest hurdles to what would turn out to be the world's largest computation were not technical, but political. The U.S. Government had engaged in a three-year criminal investigation of Phil Zimmermann after a cryptography program he wrote called Pretty Good Privacy (PGP) found its way onto the Internet and out of the country in 1991.

The government claimed that its export rules for cryptographic products had been violated and threatened to prosecute Zimmermann as a dealer of illegal munitions. Although the investigation had been officially dropped in early 1996, there was still no precedent—no one had been known to export cryptography from the U.S., so anyone trying would likely be the test case to see whether the government would actually attempt to prosecute violations of cryptographic export policy.

Rocke Verser did not like the prospect of becoming the subject of a government investigation, so he decided to limit his project to citizens of the U.S. and Canada working in the U.S. and Canada.

Given Zimmermann's hassles, restricting access to the clients was a prudent move for Verser, but it did have some pretty significant consequences for the DESCHALL project. Although the U.S. had played an important role in the development of computing and the Internet, the technology wasn't confined to North America. It had been many years since the U.S. and Canada contained more processing power than could be found in the rest of the world combined. People outside the U.S. who wanted to participate in RSA's contest would need to find another way.

In an article posted to the DESCHALL mailing list, Verser called for an attorney well-versed in the law surrounding the issue of cryptography, specifically its export from the U.S. to advise us. Although there was interest, we could not find a lawyer who could help us.

Another option that was considered was the publication of the source code for Verser's clients. Under the law of International Tariffs in Arms Regulations (ITAR), descriptions of mechanisms were not forbidden, but implementations were not allowed outside of the U.S., except to Canada. Thus, source code, printed on a piece of paper as a description of the system could be exported, but the same software on a floppy disk, as an implementation, could not. The problem with this approach was that it would simply take too long to prepare software for publishing, to have the printing and shipping done, and for the recipients to run the paper through the necessary optical character recognition software needed to turn the printed source code into working software.

What we did not know at the time was that the Clinton administration moved cryptographic software from coverage under ITAR three months earlier. We had no idea that William A. Reinsch from the Department of Commerce described regulatory policy on cryptography just nine days earlier for the House Judiciary Subcommittee on Courts and Intellectual Property. Even if we had known that cryptography was now regulated under the Bureau of Export Administration (BXA), part of the Department of Commerce, as a "dual-use" technology—that with military and non-military uses, the rules were so new, we wouldn't have known how to follow them anyway.

The final and most feasible suggestion was to have clients developed outside of the U.S. work with our project. However, this software did not exist and as was becoming clear by demonstration in the European DES-Challenge effort, unless an idea had software to back it up, nothing more would come of it.

The core DESCHALL group had other problems to address, namely barriers that prevented machines in the U.S. from participating. There were two general classes of such machines: systems that were connected by modems (and were thus often offline) and systems whose Internet connections passed through restrictive network firewall systems.

The occasionally-connected were basically all of the individual machines: home computers that were used for e-mail, the Web, and games. Without constant connectivity to the Internet, the DES key search software would not be able to communicate with the keyserver in a timely fashion. The keyserver gave out keys in packages that took the slower computers about forty-five minutes to test. Once the computer finished a batch of keys, it needed to reconnect to the keyserver in order to report its results and get a new batch of keys to test. For most people whose computers shared the household telephone line, this was an inconvenience. The problems facing these users would be difficult to address, so we began with an easier problem: getting the systems behind firewalls able to participate.

Computers whose Internet connections were regulated by network firewalls—millions of them throughout North America—had tremendous computing power and could easily form the basis for the world's largest virtual computer, if only we could harness their power by getting their managers to put the DESCHALL client software on them. The simple fact that most of these machines ran behind firewalls prevented them from being able to talk to our keyserver.

The most straightforward way to connect to corporate clients to our keyserver would be to take advantage of the ever-growing ubiquity of the Web. Simply stated, we needed clients to be able to talk to the keyserver via the Web's protocol: HTTP, the HyperText Transfer Protocol. HTTP was basically a message format that would specify how to format a communication between a Web server and a Web client (like a browser). Many corporate networks already had made the necessary changes to allow HTTP traffic through their firewalls safely. The challenge facing us there was how to make that happen without introducing more code into the clients—which needed to be as small and unobtrusive as possible—and without imposing new requirements on the keyserver.

Tuesday, April 1, 9:35 P.M.
Worcester Polytechnic Institute, Worcester, Massachusetts

From his dorm room at WPI, student Carleton Jillson was looking at Rocke Verser's Web site and was studying some of the basic statistics that had been made available. Those statistics showed that eight of the top ten contributors, in terms of keys tested, were educational institutions. Quite a lot of additional processing power was potentially available from similar sources, and university systems could run DESCHALL as is, without any additional modifications needed for things like working through firewalls.

To address the needs of the computer lab managers who wanted to participate with all of the machines at their disposal, we would need to provide the software not only to test keys, but to manage the key-testing software automatically on the dozens or hundreds of machines that a lab would contain. Making these processes automated would help the lab managers contribute a great deal of computational power without a lot of time and effort.

Something else that would help to inspire lab managers to contribute their systems' idle cycles would be to let them see just how much processing power they had at their disposal. If we let them see how many keys they were testing, we might find that various lab managers would start to compare their results with others, using some friendly rivalry to induce them to throw still more processing power at DESCHALL.

At this point in the project, we had searched less than two tenths of one percent of the available key space, which despite being a long way from the conclusion, was far ahead of the other efforts. The progress of all three public efforts is summarized in Table 2. Jillson sent a note to the DESCHALL mailing list with his observations on our relative standing in the contest.

Group	*Keyspace Completed*	*Millions of Keys Tested per second*	*Average Time to Find Key*
DESCHALL	0.162%	165.000	7 years
SolNET	0.013%	50.704	23 years
DES Violation	0.058%	43.000	27 years

Table 2. Status of DES Challenge Groups

Shortly after Jillson wrote his thoughts to the DESCHALL mailing list, I read his message and reflected on our standing. Not having seen any traffic on the European DES-Challenge mailing lists for about two weeks, it was becoming clear to me that the group that won RSA's 48-bit RC5 contest was simply not going to write the software needed to solve RSA's DES Challenge.

Thinking about the comparison among the three active public projects, I was pretty happy to be working with DESCHALL. We were clearly the front-runner in the contest, and our rate of 165 million keys per second seemed impressive enough. It certainly *sounds* like a lot of keys. A more useful, and sobering statistic, was the time needed to find the right key at our current testing rate—seven years. If we were to succeed, we needed to get a lot more people to participate in DESCHALL.

10

Needle in a Haystack

Thursday, April 3, 4:01 P.M.
The Ohio State University, Columbus, Ohio

Justin Dolske sat in his lab and and tried to think of ways that he could encourage more people to participate in DESCHALL. Dolske knew that we needed to find a way to describe the nature of the problem we were attacking.

Looking for the right key out of all possible DES keys is a big job, basically the equivalent searching for a needle in a haystack of 72 quadrillion strands. Dolske grinned as the obvious question presented itself: "How big is the haystack?"

"Figure that a strand of hay is a cylinder ten centimeters long and two millimeters wide," typed Dolske. "Then assume that the hay is packed densely. Finally, let's assume that a haystack is shaped roughly like a sphere cut in half. After crunching the numbers we see that our haystack is roughly two and a half miles wide and over a mile high."

Dolske hit the "send" button and shot a copy of his observation to the DESCHALL mailing list for other participants to see.

Friday, April 4
Carnegie Mellon University, Pittsburgh, Pennsylvania

CMU graduate student Bridget Spitznagel updated her Web document, Frequently Asked Questions about DESCHALL and the DESCHALL effort at CMU. She thought putting the magnitude of our problem in monetary terms might be fun.

For her calculations, Spitznagel assumed that U.S. paper currency bills are six inches wide, two and a half inches tall, and one one-hundredth of an inch thick. If each possible key were worth a penny, the entire key space would amount to one square mile of $100 bills that was twenty-two feet thick.

Put another way, if potential keys were pennies, we'd be looking for one penny out of over $720,575,900,000,000 worth of pennies.

Friday, April 18, 5:01 P.M.
Virginia Polytechnic Institute, Blacksburg, Virginia

After reading some of the messages posted to the DESCHALL mailing list two weeks earlier, computer science undergraduate student Alex Bischoff started thinking about keys. He wondered, "What if cryptographic keys were like keys for door and car locks?" An image of mountains of little metal keys suddenly popped into his head.

Then he started some calculations. Assuming that such keys are two inches long, if you laid them end to end, you'd have a line of keys long enough to circle the sun 3894 times.

Visualizing just how many combinations we would need to try gave us pause. To try so many possible keys, our DES key-cracking system was going to need a lot more clients, since they would be doing the real work of the DESCHALL project.

11

Spreading the Word

Tuesday, April 1, 10:15 P.M.
Megasoft Online, Columbus Ohio

I continued thinking about the challenge before us after reading Carleton Jillson's message. If the way to defeat DES was to get more key-cracking clients running, we needed to let a lot more people know about the DESCHALL project and to convince them to run our client software. We had to find the right people and we needed a compelling message to get their attention.

Building on that initial awareness would be the hard part. We were all pretty sure that once things got started, we could get some critical mass of participants and then wait for one of the clients to find the right key. We didn't know just what would constitute critical mass, but we knew that we were nowhere near it. At the rate we were going, we would take eight years to find a DES key. We needed thousands of clients—that would mean hundreds or even thousands of new participants.

To bring our message to a large number of people, we looked at the media, with particular emphasis on the news outlets that were reporting on computing technology. Early conversations with writers in the media were helpful. Once they understood what we were doing and why anyone would want to find DES keys, they often expressed interest in our project and wanted to be advised in the event of any major milestone (in particular, once someone found the right key). Through those conversations we learned that we didn't have time to educate people about cryptography, how DES was used, and cryptographic export policy. Reporters need to know what happened so they can give their readers the facts. We quickly learned to adapt our message to get their attention first and to fill in the details afterward. A typical story pitch might go something like, "The government standard for cryptography, used to protect the nation's financial systems is vulnerable to attack. I think your readers might like to know how a group of researchers, engineers, and students are using their computers to demonstrate how weak it is." With that as a basis, many reporters would want to hear more.

Not all DESCHALL participants were talking to reporters, though. Some of us were simply looking for ways to raise awareness among people we encountered in our daily online activities. Many of the DESCHALL participants were active on a system called *Usenet.* Usenet works much like e-mail, except that instead of being a one-to-one communication mechanism, Usenet is many-to-many. Instead of writing an article and addressing it to a person, authors will address it to a *newsgroup*, and servers all around the world will carry that article in that newsgroup. Thus, people all over the world with similar interests can read articles that people have written and post their own for others to read. Usenet would prove to be an effective way for DESCHALL participants to draw attention to the project.

Signature blocks have long been a part of Usenet articles and e-mail. The basic idea is to define some block of text that will be automatically appended to your message, rather than making you retype your name on each message. Before long, people started adding more information to the signature block, including contact information, thus creating a sort of virtual business card. Pithy remarks were also included on occasion, and some people even went so far as to create huge signature blocks, with gaudy pictures made out of text characters, advocating a dozen different causes. Taken to this extreme, signature blocks could become the electronic equivalent of the bumper sticker.

Before long, messages showing up on the DESCHALL mailing list were carrying signature blocks that advertised the project or provided a link to the project Web site. As DESCHALL project participants went about their business, their signature blocks advertising DESCHALL started to spread. Usenet newsgroups, e-mail lists, and private correspondence became graced with mentions of and links to DESCHALL, usually with a simple tag like "Crack DES Now!" (Although we weren't technically attempting to crack DES itself—we were trying to crack a DES-encrypted message—our experiences with the media helped us to understand that opening with a long technical digression would not catch and hold the reader's attention. Brevity rules.)

Likewise, on their personal Web sites, participants began to describe the project and their efforts to advance it. Invitations to join the project were often extended on such Web pages. Oregon State University engineering student Adam Haberlach and I made small graphical buttons fashioned after the "Netscape Now!" buttons that graced so many Web pages in 1997. In a problem akin to having a cupholder with no car to put it in, the European DES-Challenge group that never made any software had created a Web site and graphics. One particularly common graphic was a "Crack DES Now!" button that came from that group. Justin Dolske commandeered that button and put a copy on his Web site for others to use. Since the European DES-Challenge effort had no software, it didn't seem that they would need the promotional graphic themselves.

Dolske didn't really have time to try to create new graphics of his own. He had been drafting a "call for participation" document with a brief description of the project and its purpose which was aimed at the technically inclined who would be most likely to understand the project without any explanation. Dolske's call was posted to Usenet where it would be seen by others involved in cryptography.

The increasing mentions of DES and DESCHALL online helped us recruit new participants who, in turn, encouraged others to join DESCHALL.

Thursday, April 3, 2:30 P.M.
Megasoft Online, Columbus, Ohio

A critical aspect of the promotional effort was to stress the importance of the DESCHALL project to others who weren't cryptographers and might not even use computers much themselves. To find a way to relate DES security to the concerns of a typical American citizen, I called my own bank, KeyBank, introduced myself as a computer scientist working on a security research project, and asked to speak with someone in the bank's information security group. The person who answered the phone took my name and number, promising to have someone call me back. Shortly thereafter, my call was returned, and the bank representative and I engaged in an interesting discussion about cryptography, specifically the use of DES. Although the bank official did not want to share details of how DES was used in the banking industry, he was willing to verify for the record certain vague statements like "DES is heavily used in the financial sector." He expressed serious interest in the project, wished us success, and said that he would be watching our progress from "a safe distance."

Tuesday, April 8, 7:22 A.M.
Loveland, Colorado

Among the hats that Rocke Verser wore throughout the day was that of editor. Justin Dolske and I worked with Verser to create a press release that would help more DESCHALL participants to talk to the media with confidence. Draft after draft, the press release got improved. Finally, Justin Dolske, Rocke Verser, and I had something we were reasonably happy to share with the rest of the project participants.

Many of the newcomers to the project were very enthusiastic, but did not have the kind of background in cryptography needed to frame the discussion in the right context for reporters on their own. Part of the motivation for our press release was to provide the less technical participants with a simple fact sheet that would help them to make the pitch to their local media outlets. Once the release was put on my Web site and posted to the DESCHALL mailing list, participants began calling local media, pitching a story about the project, with a connection that would be of interest to local news organizations—someone from the immediate community participating in a nation-wide effort.

Hoping that if we addressed tech-savvy media would help us find still more participants, I sent a draft of our press release to the tips contact address at News.com.

1:30 P.M.
CNET Networks, San Francisco, California

Courtney Macavinta, a writer at CNET's News.com found the announcement of the DESCHALL group's formation of interest. Given the success of the recent 48-bit and 40-bit challenges, she thought that DESCHALL might actually have a shot at solving the challenge.

After reading the press release, she telephoned Rocke Verser and tracked down a few more sources that could help to estimate the difficulty of the problem. She finished her article, and it went into the News.com publication system.

"Users take crack at 56-bit crypto" ran on News.com with a lead-in that clearly set forth the seriousness of our claim, as well as the difficulty facing us. Macavinta wrote,

> Thousands of American and Canadian computer users are working night and day to prove that the 56-bit encryption standard set by the United States government is vulnerable. But the effort could take several years.

Our objective was to draw some more attention to the project, bringing in a whole new audience of potential participants. Articles like the one that CNET ran were critical in these efforts.

Wednesday, April 9, 5:50 A.M.
Megasoft Online, Columbus, Ohio

Happy to see the success with CNET, I sent a copy of the press release to my local paper, the *Columbus Dispatch*. After a long night of working on DESCHALL, I posted a copy of the press release to the DESCHALL Web site that I maintained.

> DESCHALL GROUP SEARCHES FOR DES KEY
> Sets out to prove that one of the world's most popular encryption algorithms is no longer secure.
>
> COLUMBUS, OH (April 9, 1997). In answer to RSA Data Security, Inc.'s "Secret Key Challenge," a group of students, hobbyists, and professionals of all varieties is looking for a needle in a haystack 2.5 miles wide and 1 mile high. The "needle" is the cryptographic key used to encrypt a given message, and the "haystack" is the huge pile of possible keys: 72,057,594,037,927,936 (that's over 72 quadrillion) of them.
>
> The point? To prove that the DES algorithm—which is widely used in the financial community and elsewhere—is not strong enough to provide protection from attackers. We believe that computing technology is sufficiently advanced that a "brute-force" search for such a key is feasible using only the spare cycles of general purpose computing equipment, and as a result, unless much larger "keys" are used, the security provided

by cryptosystems is minimal. Conceptually, a cryptographic key bears many similarities to the key of a typical lock. A long key has more possible combinations of notches than a short key. With a very short key, it might even be feasible to try every possible combination of notches in order to find a key that matches a given lock. In a cryptographic system, keys are measured in length of bits, rather than notches, but the principle is the same: unless a long enough key is used, computers can be used to figure out every possible combination until the correct one is found.

In an electronic world, cryptography is how both individuals and organizations keep things that need to be private from becoming public knowledge. Whether it's a private conversation or an electronic funds transfer between two financial institutions, cryptography is what keeps the details of the data exchange private. It has often been openly suggested that the US Government's DES (Data Encryption Standard) algorithm's 56-bit key size is insufficient for protecting information from either a funded attack, or a large-scale coordinated attack, where large numbers of computers are used to figure out the text of the message by brute force in their idle time: that is, trying every possible combination.

Success in finding the correct key will prove that DES is not strong enough to provide any real level of security, and win the first person to report the correct solution to RSA $10,000.

Many more participants are sought in order to speed up the search. The free client software (available for nearly every popular computer type, with more on the way) is available through the Web site. One simply needs to follow the download instructions to obtain a copy of the software. Once this has been done, the client simply needs to be started, and allowed to run in the background. During unused cycles, the computer will work its way through the DES keyspace, until some computer cooperating in the effort finds the answer.

If you can participate yourself, we urge you to do so. In any case, please make those you know aware of our effort, so that they might be able to participate. Every little bit helps, and we need all the clients we can get to help us quickly provide an answer to RSA's challenge.

With the CNET article published and a press release on the Web site, my workday of over twenty-four hours came to an end.

After a few hours' sleep, I was back online, watching the mailing list, seeing other participants describe their efforts to get more publicity for DESCHALL. All told, local papers in Minnesota, Michigan, Ohio, Connecticut, California, and Ottowa were contacted by participants in those areas. Some participants contacted the national technology media and broadcast media throughout the United States and Canada. It was a busy day.

In the first half of 1997, few in the mainstream media understood the significance of the Internet, what kinds of possibilities it presented, or even why anyone should care about DES. A larger problem was that, while most reporters were interested, they didn't really see a story in the beginning of an effort. If we managed to succeed, however, they wanted to hear about it.

This reaction was not altogether surprising, but it was frustrating in light of our early success with CNET. We were very happy with the coverage that we did get—even if only CNET picked it up. Thanks to that one article, we got the attention of new participants, which is just what we needed—even if it wasn't the worldwide mainstream media coverage we wanted.

Thursday, April 10, 1:39 A.M.
The Ohio State University, Columbus, Ohio

Justin Dolske looked over RSA's Web site, and its description of its 1997 Secret Key Challenge. Noticing a link called "In the News" for the first time, he clicked on the text. Dolske noted the links to the articles written about RSA's 40-bit and 48-bit challenges being won. In addition, he saw a link he did not expect to find: one to CNET's April 8 article.

Dolske smiled and fired off a message to the DESCHALL mailing list. Attracting enough attention for the contest sponsors to notice us would be important, because anyone finding out about the challenges from RSA's site would be able to follow links to see that RSA's DES Challenge was being answered.

"Nice to see that RSA knows that they may need to get out their checkbook soon," observed Dolske in his e-mail.

As the days went on, we realized that our approach of a simple press release that individual participants would use to base their own pitches to local media was a good one. Rather than having a single Associated Press story (for example) that everyone would run, each paper got to write its own story about someone from among the readership that was involved in a very important project dealing with the security of cryptosystems. The press release provided the necessary background and the rest of the story was about the involvement and the trials of the local individuals participating.

This strategy was at its most effect when the press release was sent to university newspapers. Many students pitched stories to their school papers, and, taking a cue from Carleton Jillson's April 1 message to the mailing list, would point out their standings in comparison to rival schools.

12

The Race Is On

The way we in the DESCHALL project saw it, friends didn't let friends have idle computers. This attitude helped us recruit as many participants as our publicity efforts did, perhaps even more, and this sort of informal recruitment was particularly prevalent on college campuses. Most of our processing power was coming from universities—not really a surprise, given the kind of cultural differences between corporations that wanted to reduce complexity on their production systems and the comparatively freewheeling universities where people often run programs for no other reason than that they could. Further driving the trend for participation from college campuses was the simple fact that most students had their own machines in their dorm rooms, and many large universities provided network to the campus network to dorms.

Students at Worcester Polytechnic Institute (WPI) in Massachusetts managed to work their way to second place in the per-domain rankings by running the DESCHALL clients on their own personal computers in the middle of March. The twenty-four machines that were running DESCHALL were processing more than 784 billion keys per day. The Institute's computer lab managers had banned the use of our clients on their lab machines, so WPI students enlisted the help of their friends as well as their own personal computers. As the weeks wore on, WPI students would not be able to keep up with the key testing rates at other universities.

Even in early April the processing power that we had harnessed at universities was massive. On April 8, for example, DESCHALL tested a total of 24 trillion keys. That was a rate of 277 million *each second*, for every second, around the clock in that single day. That rate was roughly *ten times* the rate of Ian Goldberg's answer to the 40-bit Challenge—

but still just over half the speed of Germano Caronni's 48-bit Challenge project.

Statistical analysis of our key-testing rates was critical, since participants wanted to be able to see how the project was progressing overall. Of more interest to many participants was the breakdown showing each participating "domain"—the group of machines in each organization's online name, such as ohio-state.edu or megasoft.com.

Looking at the per-domain statistics allowed participants to see how much they were contributing by comparison to other organizations. This turned out to be an excellent way to foster some healthy competition, particularly among universities where rivalries had developed over the years. Table 3 shows the top participating domains for April 8.

Keys Tested	*Clients*	*Contributor*
4.8 Trillion	278	Oregon State University
2.3 Trillion	182	Rensselaer Polytechnic Institute
1.4 Trillion	25	Rochester Institute of Technology
1.3 Trillion	40	Worcester Polytechnic Institute
1.3 Trillion	196	Ohio State University

Table 3. April 8, Top Contributors per Domain

Wednesday, April 2
Oregon State University, Corvallis, Oregon

Unlike WPI, lab personnel at Oregon State allowed students to run the DESCHALL clients on machines in the public computer labs. Oregon State managed to grab the top spot for DESCHALL key searching and to hold its title for several weeks.

An engineering student there, Adam Haberlach, was largely responsible for Oregan State's participation. Haberlach had seen a reference to DESCHALL on a mailing list for the now-defunct DES-Challenge group. He downloaded the client software in mid-March and ran it on his laptop at home. Haberlach worked in a test lab run by the Business Department there with about sixty client machines that had spent a lot of time doing nothing, so he decided to put the computers to work. When Haberlach got to work the next day, he promptly installed the DESCHALL client onto his computer. Later that morning, he managed to persuade one of his coworkers to install the client on his machine. As word of the DESCHALL project spread trough Haberlach's office, more and more employees installed the client until nearly all of Haberlach's coworkers were participating.

Even after harnessing all of this power, Haberlach wasn't finished. In the same building, at the other end of the hallway was a lab with another 160 machines. Haberlach was eager to install DESCHALL on all of these machines, because he knew that spring break was imminent and soon these machines would be spending all of their time running screen savers. Haberlach approached the management of the larger lab about running DESCHALL on its machines. Haberlach explained the importance of DES and the prestige the University might gain from participating, particularly if they contributed a substantial amount of computing power or if they found the key. The lab management were impressed by Haberlach's arguments and gave him permission to run DESCHALL on all of the machines in the lab.

Within thirty minutes of the lab being closed, Haberlach and his group had all of the machines running DESCHALL. Having seen the impact of these machines on the project overall and the role they played in taking Oregon State to first place, lab management started approaching other lab administrators and trying to drum up more support for the project and for Oregon State's ranking. By the time DES fell, Oregon State had tested over six trillion keys, making it one of the top ten institutions in terms of the number of keys tested.

While Haberlach and other DESCHALL enthusiasts were rapidly increasing participation at Oregan State, others were developing new clients that would allow more people across the country to contribute to our efforts. Several participants had developed programs for Unix machines that would search for DES keys when the machines were idle and then pause this search when someone was using the machine. The end result would be that people who needed to use the computer would not need to share any of their system's resources with a piece of software like DESCHALL, and that when these computers were not being used, their spare CPU cycles could contribute large amounts of effort to the project overall.

Friday, April 11
Megasoft Online, Columbus, Ohio

I looked over our statistics to see just how much progress we were making. Our statistical reporting was simple, but it got the job done. Each day, we reported the number of keys that we tested that day and the running total of keys tested. Recognizing the psychological difficulty of running a race without knowing how far away we were from the finish line, we had to find a way to represent the goal for our participants. That need was satisfied by another statistic that some people found curious—"Time to 50%."

We didn't want to make the goal seem too far out, and putting the time needed to search the entire keyspace would be an unrealistically long period of time. We had only one chance in over 72 quadrillion of having to search the entire keyspace. On the other hand, we didn't want the goal to seem too unrealistically close. In the end, Rocke Verser decided simply to use the mid-point as the goal we watched because on average, any brute force attack would need to run through half of the total keyspace. In any case, it was a useful metric, allowing participants to think about how long it would take a project like DESCHALL to crack not just one key, but DES keys in general, based on the key testing rate that DESCHALL had achieved.

Looking over the statistics showed us several important milestones, and how far we had managed to come in just under a month. Given DESCHALL's performance in mid-March, we were seventy-five years away from finding the DES key we wanted. Organizing the project in March gave us a boost, bringing our search time down to under ten years. Within a few days of the mailing list being brought online, we managed to cut that time in half. With more participants joining the project in the past two days, we had managed to pass the point of having searched roughly half of a percent of the total keyspace. The day before, we managed to sustain a key search rate of 300 million keys per second.

Looking over these statistics, I could see that about March 21, the project started to increase its rate of progress dramatically. Seeing that this is when we started to get some of the labs of systems sitting idle at universities, I thought this confirmed our strategy of making the software as easy to run as possible. In addition to making the key-cracking client easy to start, having it operate quietly in the "background" would

Date	Daily Keys Tested	Total Keys Tested	Estimated Time to 50%
03/14/97	1313404354560	14017854701568	75.0 years
03/15/97	2342485229568	16360339931136	42.0 years
03/16/97	2881570734080	19241910665216	34.0 years
03/17/97	2560471597056	21802382262272	39.0 years
03/18/97	2780831940608	24583214202880	35.0 years
03/19/97	2997157363712	27580371566592	33.0 years
03/20/97	3724239962112	31304611528704	26.0 years
03/21/97	6298259161088	37602870689792	16.0 years
03/22/97	9185290878976	46788161568768	11.0 years
03/23/97	9667770056704	56455931625472	10.0 years
03/24/97	8623505801216	65079437426688	11.0 years
03/25/97	7939918135296	73019355561984	12.0 years
03/26/97	9066726293504	82086081855488	11.0 years
03/27/97	10519398318080	92605480173568	9.0 years
03/28/97	10115017080832	102720497254400	10.0 years
03/29/97	14256128917504	116976626171904	7.0 years
03/30/97	14365398925312	131342025097216	7.0 years
03/31/97	12454079758336	143796104855552	8.0 years
04/01/97	12892275474432	156688380329984	8.0 years
04/02/97	15425928691712	172114309021696	6.0 years
04/03/97	14122640998400	186236950020096	7.0 years
04/04/97	17942393651200	204179343671296	5.5 years
04/05/97	22777738297344	226957081968640	4.3 years
04/06/97	21748321878016	248705403846656	4.3 years
04/07/97	21215989202944	269921393049600	4.6 years
04/08/97	23947613569824	293869006619424	4.1 years
04/09/97	25581437582560	319450444201984	3.9 years
04/10/97	28855494508544	348305938710528	3.4 years

Table 4. DESCHALL Statistics Through April 10

prevent many from not running DESCHALL because they believed it somehow interfered with a computer's normal day-to-day functions.

It has been said that it's easier to get forgiveness than permission, and while this may be true for a one-time event, our early experiences indicated quite the opposite is true in an ongoing effort such as ours. Dealing with computing management worked better than letting managers discover how much of their systems' computing power was being devoted to the project by stumbling across it. Likewise, getting cooperation from the policy makers by explaining the project, its importance, and, perhaps most important, showing how the software didn't interfere with normal work proved itself effective. Not only were we getting permission for the specific networks in question, but we were winning advocates to the DESCHALL cause.

13

Clients

Those statistics showed us that we had come a long way since March. The problem was that we were still far less than one percent through the keyspace, and we could expect that we were years away from finding the key. We knew that we needed to make the DESCHALL project run faster—to test more keys in a day. We already had enthusiastic participants helping to get more systems running the DESCHALL client software. We needed more speed, and that would have to come not only from getting more systems working on the project but in getting that client software to run faster. How can we get the software to run faster when the DES algorithm itself specifies each step that must be taken to take ciphertext and to turn it into the corresponding plaintext?

The first insight into the process of testing keys quickly is that you don't need to try them one at a time. One of the beauties of a procedure with many steps is that by putting the steps in a particular order, you can reduce the number of steps overall. To illustrate, suppose you are packing your car for an overnight trip; you don't take each article of clothing out of a drawer all the way outside to your car, open the trunk, place it inside, close the trunk, then back for the next item. Instead, you determine what you need, place everything into bags—perhaps an overnight bag and a garment bag—thus reducing the number of trips from the house to the car from the number of items you're taking to two.

A further refinement could occur if you observe that the process of opening and closing the trunk works the same way for each bag. You'd notice that you are simply filling a bag, taking it to the car, opening the trunk, placing the bag inside, closing the trunk, and then going back to pack another bag. Once you see this and recognize that the process of putting a bag in the trunk is exactly the same for each bag, you could pack both bags, carry one in each hand to the car, open the trunk, place both bags inside, and then close the trunk.

Just as anyone who has ever packed for a trip can order and reorder events to reduce the amount of time and effort needed, programmers can increase their systems' efficiency by studying a particular process and finding ways to eliminate redundant work, a strategy called *optimization.*

This is precisely what Rocke Verser did with our DESCHALL clients. By studying DES, he was able to find ways to perform DES encryption and decryption very quickly. By studying DES decryption in greater detail, he found ways to test for valid keys even more efficiently than using his fast decryption processes.

January 1997
Loveland, Colorado

While looking over the process of decrypting DES ciphertext, Rocke Verser got an idea. There was no need to program the computer to execute the decryption routine in its entirety with one key, then to execute the entire decryption routine with the next key, and so on. By looking at each step in the decryption process individually, Verser could find ways to save himself some effort in searching lots of DES keys.

Verser watched the results of the first steps of a DES decryption with each key with which he performed that calculation. He quickly noticed that each DES key has another DES key that is mathematically related to it—a *complement*—that behaves identically in the first few steps of the DES decryption process. A complementary key is like a photographic "negative." Although the colors are backward, the first few steps in identifying the subject of the photo (finding the outlines) will be the same for both the photographic print and the negative.

Verser could take advantage of this property so that instead of performing those initial steps for every single key to be tested, he could perform those steps once—for both a key and its complement at the same time. Since this worked only for the first few steps of the decryption process, it didn't cut the number of decryptions in half, but it did reduce the amount of work needed to test two keys.

After further study, Verser found that a key could be identified as incorrect for the ciphertext to be decrypted well before the entire decryption process would run its course. In the DESCHALL key-testing client, Verser wrote four tests to find when a key was the wrong one so he could make his DES key testing client run even more quickly. The vast majority of incorrect keys would be identified by one of the first two tests—only one out of 4096 keys passed both of these tests.

After running a few more steps through the decryption process, the DESCHALL client would run a third test, after which only one key out of about four billion would still be a possible match. Verser called a key passing this test a "half-match" because half of the bits in the key were correct, like having half of the numbers in a 56-number combination.

The DESCHALL client then performed another step in the decryption operation and then one more test. If the result of the test is what's expected, the key is a "full-match," which almost certainly makes it the correct key.

Between the early-stage reductions in processing and the late-stage detection for non-matching keys, Verser's method for testing DES keys was dramatically more efficient than running each key straight through the decryption process. DES decryption straight through takes sixteen rounds—Verser's method tested keys in under twelve. These performance improvements allowed for many more keys to be tested each second.

Verser knew that his key-testing program would need to run on many different kinds of computers, from the engineering workstations found at universities and in scientific institutions to desktop PCs. Even among the PCs there would be considerable variation, since Windows, OS/2, Linux, FreeBSD, and other operating systems would need their own versions of the software.

The C programming language was well-suited to the task. Originally written in 1973 at AT&T Bell Laboratories by Brian Kernighan (pronounced "Kern-ih-han") and Dennis Ritchie for the purpose of building the Unix operating system, the language had two important properties that mattered to Verser: it was fast and it was *portable*, meaning that its code would be able to require very little effort to build on different computer types. A computer type—defined as a particular combination of hardware and operating system—is known as a *platform*.

Verser built his software in C and created clients for some of the most popular platforms of the day—including Windows 95 on Intel, Solaris on SPARC, and IRIX on MIPS. He had another trick up his sleeve, though, for platforms that used Intel microprocessors. Now, C code is fast, but not nearly as fast as Assembler. (See page 60 for discussion.) With the kind of detailed control over the chip's operation offered by Assembler, Verser was able to provide precise instructions to the processor so it could have its calculations organized so they could be performed very rapidly.

Especially with the Pentium and Pentium Pro processors, Verser's code was able to run at a phenomenal speed, allowing modest desktop computers with Intel processors to run circles around $20,000 scientific workstations.

Friday, April 4, 9:12 P.M.
Health Networks Australia, Brisbane, Australia

System administrator Andrew Glazebrook posted a message to the DESCHALL mailing list from Down Under—not itself particularly strange, because there was no restriction on who could join the mailing list. Without drawing attention to where he was from, Glazebrook's message showed that, despite our limited-access client distribution system meant to comply with cryptographic export policy, the DESCHALL client had found its way to Australia.

Glazebrook was running the DESCHALL client on three of his machines, a 90 MHz Intel Pentium running OS/2, a 66 MHz Intel 80486-DX2 running Linux, and another Linux system, with a 133 MHz "Pentium compatible" processor. Virtually all software—including operating systems, office suites, and Web browsers—was written in a high-level language (like C++) and then turned into machine code by a compiler for execution on the computer. Because the "compatible" processors had the same instruction sets as the genuine Intel processor, they could execute the same machine code.

Even though the processors from Intel and the compatible processors made by other manufacturers (like Cyrix and AMD) had the same instructions, they had very different internal structure—each manufacturer's chip was wired differently to avoid infringing on Intel's intellectual property. Usually, the difference in internal structure made no difference to the user—software produced by a compiler would run about the same speed on a processor from Intel or another manufacturer.

Rocke Verser's key-testing software for Intel processors was not produced by a compiler, however. Because of all of the optimization that the DESCHALL client had for various Intel processors, the differences in processor internals made a huge difference. Much to Glazebrook's dismay, his 133 MHz Intel-compatible system ran the DESCHALL client much slower than his 90 MHz Intel system—at a speed more comparable to his 66 MHz Intel system.

Glazebrook's message to the DESCHALL mailing list was to see if other participants were having similar experiences with Intel compatible processors. They were, and calls for clients built for Intel-compatible processors started to flow in.

April 9, 5:14 P.M.
Loveland, Colorado

Rocke Verser made an announcement on the DESCHALL mailing list: Justin Dolske and Guy Albertelli, another graduate student at Ohio State, were granted access to the source code and would soon begin work on porting the DESCHALL client software to work on other systems. Verser had been looking for ways to get his DESCHALL clients running on other platforms—such as the hot graphics-oriented machines from Silicon Graphics, Inc. (SGI), the sophisticated network-centric systems from Sun Microsystems, the datacenter-friendly RS/6000 systems from IBM, and the lightning-fast DEC Alpha systems. Verser hoped that with more platforms able to run the client, we could get greater numbers of participants, in turn helping us to find the key more quickly.

Fig. 4. Justin Dolske, Matt Curtin, and Guy Albertelli at the Ohio State University, 1997

In a period of a few weeks, the expanded development team produced a series of new clients. SGI machines running IRIX 6.2 on MIPS processors got their own client. IBM's RS/6000, running AIX 3.2, got its own client, as did AIX 4.1-based RS/6000s. Sun's SuperSPARC processors under Solaris got a client of their own. Additional clients to follow were for Digital Unix on the fast 64-bit DEC Alpha processor. Finally, Sun's Solaris on the Intel processors got a client.

While Dolske and Albertelli worked on the clients, I continued thinking about the issue of connectivity. Getting those systems behind firewalls to be able to run DESCHALL clients would require a good look at our overall architecture.

14

Architecture

DESCHALL's architecture—the electronic infrastructure to support the testing of DES keys—was a simple one. As Rocke Verser originally designed the DESCHALL system, there were only two components: the key-testing client software that project participants would run on their computers and the keyserver that would keep the clients coordinated, making sure that each DES key was being tested—and tested only once.

The DESCHALL keyserver was located in Verser's home office in Loveland, Colorado, about fifty miles north of Denver. As clients requested blocks of keys to test, the keyserver would assign those blocks of keys, keeping track of which blocks were assigned and when. This keyserver would keep clients from wasting their efforts by testing keys that other clients already tested.

This is how coordinated efforts like SolNET, DES Violation Group, and DESCHALL differed from an uncoordinated key search system like the DESKR software Peter Trei wrote. DESKR was called an uncoordinated key search because each client would randomly pick DES keys and test them; no two DESKR clients would coordinate their efforts with each other. So, Trei created his DESKR software and made it available for people to run if they chose, but he had no "project" to manage and no way to know how much work was being done by DESKR clients that people were running. SolNET, DES Violation Group and DESCHALL all operated as projects independent of one another, but each with its own architecture of key testing clients coordinated by a keyserver.

Like the other coordinated projects, DESCHALL's architecture relied on the Internet to have clients communicate with the server. When Verser was building the DESCHALL software in January of 1997, he had to decide how to get the client and server to send their messages to each other. Adopting Germano Caronni's protocol in a February revision of the software, Verser knew what messages to send, but he still had to decide how to send them. This was a lot like being on vacation and sending a letter to a friend back home: you know what you're going to write ("Having a great time! Wish you were here!") but you have to decide whether you prefer the economy and convenience of a postcard over the cost and reliability of certified mail.

Verser wanted his DESCHALL clients to run on as many systems as possible, from all over the country. There wasn't a need for a great deal of information to be exchanged between clients and the keyserver—messages could be terse, and there weren't many possible messages that could go between the client and server. Clients needed to ask for keys, the keyserver needed to provide keys, the clients needed to report on their status, and the keyserver needed to be able to tell clients when to shut down. There simply was not much more for clients and the keyserver to say to each other.

Since many clients needed to be supported by the keyserver, Verser needed a communications protocol that was lightweight, keeping the keyserver from becoming so bogged down with unpacking messages from clients and packaging responses for the clients that it did not have time to deal with the messages themselves. To address those requirements, Verser chose to have the messages between the clients and keyserver be sent using the Unreliable Datagram Protocol (UDP).

UDP messages ("datagrams") are very simple, roughly the equivalent of scribbling a message on to a postcard. Sending the message is easy to do and they almost always reach the destination without any problem. Even so, there is no guarantee that the recipient got the message. If you want to be sure that your message has been received, you'll need to have your own method to verify the fact because the postal service won't provide proof of delivery on a postcard.

Receiving UDP messages is also very easy and like a postcard. The computer just receives the datagram and then passes it to a program for processing. The operating system does not need to count the number of datagrams to be sure that it got all of the message parts—the datagram is all there is.

An alternative to UDP is the Transport Control Protocol (TCP), which involves a lot more work to establish and to maintain the channel of communication between the client and the server. TCP is generally used to establish connections where there will be more than one tiny message sent in each direction. Designed for sending larger amounts of data, TCP can break large chunks of data into smaller "packets" that are numbered and sent separately. The computer operating system on the receiving end will then take the packets, put them into the right order, and then hand the reassembled message completely intact to the program that is listening for that message, as if it were never broken into pieces at all.

So that messages sent via TCP are always complete, TCP provides another important feature: guaranteed packet delivery. In this way, TCP is like certified mail. With certified mail, you'll need to put your message into an envelope and at the post office, they'll put an extra tag on the outside for the recipient to sign, and the carrier will not release the letter without a signature. That signature will then be sent back to the sender as proof of delivery. With TCP, the systems establish a connection, and then acknowledge receipt of each packet by number. If one computer sends a TCP packet and the receiving computer does not acknowledge the packet, the sender will resend the packet until it is acknowledged by the recipient. The computer has to do more work to send and receive TCP messages; computer systems can process far fewer TCP packets at any one time than UDP datagrams for this reason. If you need to be sure that the message is received, TCP is your best bet—unless the extra work it requires the system to do is more than your system can handle.

With either TCP or UDP, Verser would be relying on another protocol for messages to be sent from one system to another. That protocol, Internet Protocol (IP), serves essentially the same purpose for the Internet as the postal system does for global mail delivery. The postal system defines standards so that anyone can address a message to anyone else in the world unambiguously. Those standards make it possible for a message to be carried toward its destination by one mail carrier after another, until it finally arrives. IP serves the same purpose—it is literally the protocol that defines the Internet and ties the entire world together. IP defines the addressing scheme for computers' global network addresses (Internet addresses, such as 192.168.1.2) and the basic framework for how to format data for transmission on the Internet, without regard to which type of computer the recipient is using. Just as the postal addressing scheme provides foundation so that postcards, letters, and parcels can be properly delivered, IP provides the foundation that allows TCP packets, UDP datagrams, and other chunks of data to be sent from one computer system on the Internet to another.

Messages that DESCHALL clients needed to send to and receive from the server were small and could easily be contained in a single datagram: things like "Ready for keys," "Test block number...," "Finished block number...," and "Stop working." Though the messages were important, Verser decided that it was best to use UDP and then to have the DESCHALL client software simply wait for a few minutes after sending a message; if no response came back, the software would just repeat the send-and-wait procedure until it did get a response. This choice allowed clients to communicate reliably with the keyserver, while also allowing the keyserver to operate efficiently enough to support many thousands of clients at a time.

Although the original architecture envisioned by Verser had only clients and the keyserver, other needs for software arose over the course of the project. Many of these additions were provided by the participants running the clients themselves, developed for their own needs and shared with other participants through the DESCHALL mailing list.

Monday, April 7, 8:34 P.M.
Bowne Global Solutions, Piscataway, New Jersey

Unix system administrator Lee Sonko could not run DESCHALL during the day at work, but he did not intend to let that stop him from participating. Using his skills in the Bourne Shell programming language, Sonko created two programs: one to start and the other to stop the DESCHALL client. When combined with the Unix system functionality for automatic process scheduling, these programs would allow it to be run on Unix machines in Sonko's care during off-hours, while leaving the machines completely free for use during normal business hours.

The week before, Sonko posted the source code for his programs so that other Unix users would be able to have their systems automatically start and stop DESCHALL clients as needed. In his case, he ran the clients outside of the hours 5:00 A.M. to 6:00 P.M. during the week and outside of 8:00 A.M. to 5:00 P.M. on the weekends. Although the DESCHALL client was specifically designed to run with very low priority, ensuring that any other needs that the systems' owners might have would be satisfied before DESCHALL took any processing time, Sonko, like others, took the additional step of not even having the clients running when anyone might need to use the systems.

Over the course of several days, Sonko made small improvements over the original version of his "desstart" and "desstop" scripts. He felt comfortable that the improvements were worthwhile and he happily posted his new and improved programs to the DESCHALL mailing list for others to use at their own facilities.

Thursday, April 10, 6:24 A.M.
Shrewsbury, Massachusetts

Adam D. Woodbury wanted to know what the DESCHALL clients were reporting as they ran, but he couldn't watch all of the machines he was running to see what they were doing. Even if he could watch all the screens, he had better things to do than watch DESCHALL client output all day, every day. Woodbury decided to write some software that would watch his clients' operation for him and let him know whenever something interesting happened.

The first issue to address would be the need to have the clients report what they were doing in some way that could be monitored. The DESCHALL client software didn't have any function in it to create a "log file"—a file containing its periodic reports on activity (the number of keys tested, how quickly keys were tested, whether the key had been found, etc.) That kind of status was just normal program output—so if the client got started from the command-line, the output would appear on-screen. Woodbury decided to create his own DESCHALL client log file by having that program output go to a file instead of to the screen—an extremely simple procedure with his computers, which ran the Unix operating system.

Once his DESCHALL clients were writing their output to a file, those log files could be easily combined to show the activity for all of the systems. With all of the program output in one place, he could write programs that would look through the logs for things that he would want to have reported to him.

Woodbury noticed that when the client reached the end of a block of keys being searched, it would report "key not found" before requesting another block from the keyserver. This gave him an idea. He decided to have his program that read logs and reported progress alert him if the key had been found. He did not know what exactly the client would say if the key had been found. Searching the logs for the message that the key was found required knowing what to search for—what exactly the client would report in that case.

Looking at what he did see from the usual output of the program—frequent "key not found" messages as blocks of keys were tested—Woodbury assumed that the client would probably say something like "key found," or at least something about "key" without the word "not" in it. He wrote a program that would search his logs every day at 1:00 A.M. and send him e-mail showing any messages that included the word "key" without "not" somewhere else in the message.

Woodbury put his program into place shortly after he had joined the DESCHALL effort in late March and it had been running since then. When a discussion on the DESCHALL mailing list turned to what the client would report once it had found the right key, Woodbury wrote about what he had done and posted a copy of the source code for his software to search the logs, thinking that perhaps someone else would find the software useful.

10:57 A.M.
Loveland, Colorado

Rocke Verser read over the program to report messages about keys that Woodbury posted to the mailing list. Verser recognized that the program would successfully report the message that Woodbury knew he was looking for ("key found") but that it would miss several other messages that might also be useful.

In addition to the "key found" and "key not found" messages, the client could report informational messages from the keyserver that users should be able to see. Verser could, for example, configure the keyserver to check the version of the client software that a user had, and if it was particularly old or needed to be replaced for some reason, send an informational message back to the client encouraging the user to upgrade to the latest version of the client software. Verser could use the "informational message" to get any information he needed to users through the clients, which would have gotten the message as part of their response to a request.

Verser typed out a quick message to the mailing list, suggesting that instead of searching for particular messages, the opposite approach be taken. Messages that participants would recognize and know they could safely ignore should be thrown away, and all other client messages should be reported.

Saturday, April 12, 12:40 P.M.
Rutgers University, Piscataway, New Jersey

Some users were interested in seeing more detailed information from the logs than whether the key had been found. The same basic principle of creating and reading log files could be used to review client status reports on number of keys tested and aggregate them for a participant with a large number of clients to review easily. In his dorm room, Scott McIntyre had several Pentium machines running DESCHALL. He was already wondering just how much he was contributing to the project overall when he noticed some others on the mailing list talk about how many keys they were testing in a day.

McIntyre did not know how the participants with Unix clients were creating their log files, so he looked at his Windows-based client for a function to create a log. Not finding one, he thought logging might be a feature in the Unix clients not present in the Windows client, or that someone had a program to capture and to aggregate the on-screen output from the client.

Expressing his desire to see what his clients were doing like some of the participants with Unix clients were, he posted a message to the mailing list asking for ideas on how to collect the DESCHALL client output under Windows.

3:43 P.M.
Austin, Texas

Like many other participants, Carol Harris was redirecting the output of the DESCHALL client to a file to create a log that kept a record of the client's activity. Unlike most of the users who did this, she was running the DESCHALL client on a Windows system. Interestingly, she was gathering the client's output the same way that participants with Unix clients were.

The Windows operating system has a long history of borrowing features that were implemented first in other operating systems. Windows had with it a command line interpreter (usually started by running

`COMMAND` or `CMD`), which was essentially the same as the old DOS command line. Improvements made to the DOS command language over the years were made by people who understood Unix and were inspired to bring some of the Unix command language features to DOS. Redirection of program output was one of these features, thus allowing the DESCHALL client to have its output sent to a plain text file instead of appearing on the screen. That file could then be reviewed, searched, and otherwise manipulated at the user's leisure.

Harris used the Windows command interpreter on her 166 MHz Cyrix "686" machine to send the DESCHALL client's output to a file, just like was being done by some of the Unix users. Seeing McIntyre's question on the list, Harris responded, showing exactly what she typed to the command interpreter to get the desired behavior.

As more participants joined in the DESCHALL project, participants continued to help each other to overcome the obstacles they were facing. Efforts of earlier participants to manage and to monitor the DESCHALL client software also helped newcomers to get their systems participating more quickly.

Wednesday, April 9, 8:19 P.M.
University of New Brunswick, Canada

While one group of participants dealt with getting their clients to start and to stop on predetermined schedules and to report client activity, another worked on strategies for participating with only dial-up modems for Internet connectivity. Jeff Gilchrist, a student in the cooperative higher education program at the University of New Brunswick in Eastern Canada was just such a participant.

Gilchrist was running the DESCHALL client on his modest 75 MHz 486DX4 system. Running Windows 95, the system was able to test keys without any problems, but frequently had difficulty communicating with the keyserver. Gilchrist noticed that his system would finish checking a block of keys, then it would wait to talk to the keyserver. Since the system was not connected to the Internet all the time, it had to wait for Gilchrist to connect to his ISP via dial-up modem. Once online, the system would wait for several minutes before reporting its progress to the keyserver and getting the next block of keys.

As explained on page 67, clients would need to request a block of keys to test from the keyserver. Once the client finished testing the block of keys, it would need to report the results and to request another. If the computer was offline, it obviously couldn't send the message to the server. If a client needed to communicate with the server and could not, it would sit there and wait until a connection was established, thus wasting time that could otherwise be spent on testing keys.

Earlier in the day, someone using the name "Icepick" posted a message to the DESCHALL mailing list in which he mentioned a program called freeDUM, a free "dial-up manager." By running the freeDUM program, DESCHALL participants could perform their usual non-DESCHALL work, and the machine would connect to the right Internet provider on a pre-determined schedule, set to ensure that the system would be online about the time that the DESCHALL client needed to communicate with the keyserver. Once the line has gone idle for some period of time, freeDUM will hang up the line, all automatically. Configuring freeDUM to make its connections every half hour worked well, since that was about the amount of time it would take for the DESCHALL clients to work their way through a key block.

Windows users weren't the only ones trying to manage the problems of participating in the project effectively via dial-up modem.

Thursday, April 10, 10:08 A.M.
Pennsylvania State University
University Park, Pennsylvania

IBM's OS/2 operating system was well represented in the project. Not only was the keyserver itself running on OS/2, but many clients were being run on OS/2 machines. Rodney R. Korte was running some of these clients on systems whose only Internet connection was a dial-up modem.

Korte developed a program for OS/2 users with modems that would help them to avoid any wasted time waiting for an Internet connection. Implemented in IBM's REXX programming language—like the DESCHALL keyserver itself—the program would watch the DESCHALL client program's output. When it saw the "Key not found" message followed by a message that the keyserver could not be reached because the network was down, it would run a program to get the machine connected to the Internet.

Other OS/2 users began making use of Korte's program, thus making sure that their machines weren't wasting cycles that could be spent looking for keys.

10:36 P.M.
Loveland, Colorado

Rocke Verser sat at his computer, quietly reading messages posted to the DESCHALL mailing list. Heartened by the way that the spirit of the project was catching on, Verser read messages posted by participants for whom the quest to find unused computing power to harness for DESCHALL had become almost an obsession.

As accounts of how DESCHALL was brought to computer labs flowed in, a picture began to emerge. Enthusiastic participants were happy to install and run the client on the computers they could reach, but in many cases, there were obstacles to overcome.

Consider a computer lab with fifty regular desktop computers. Suppose the computers are relatively new, have 200 MHz Pentium processors, and are running Windows 95. The main obstacle to putting DESCHALL clients on these sorts of machines is that they are running Windows 95. As a result it isn't always easy to add software such as the DESCHALL client in a way that does not interfere with what a user sitting in front of the system is trying to do. Even if the program could run without generating pop-up Windows and the like, at the very least, the system wanted to clutter the taskbar with the icon for the DESCHALL client and any other software that might be running to manage any logs being created.

Computers—sometimes even entire labs of computers—like this all over the country were being used only for a regular eight-hour workday, sitting idle for the other sixteen hours. If only these systems could run DESCHALL clients outside of normal working hours, a tremendous amount of computing power could be harnessed for our project.

Thinking about how to address this problem, Verser quietly tapped on his keyboard, writing a message that would be sent to the DESCHALL mailing list.

Verser's short note simply asked if anyone had created "DESCHALL boot disks," floppy diskettes that would have just enough of an operating system to boot the system, get it connected to the local area network (which, in turn would have Internet connectivity), and then to run the DESCHALL client software. Such diskettes could easily be created to run the Linux or FreeBSD versions of the client, explained Verser. People could use their computers normally, then just before leaving the lab for the day, they could put in the DESCHALL boot disk, and reboot their system.

When the machine booted from the floppy disk, it would come on the network, start the client with the correct parameters, and start testing keys. Anyone who needed to use the machine would just eject the DESCHALL disk and reboot the machine. The system would come up in its usual environment, completely free of the DESCHALL client.

In a lab of fifty new Pentium machines, if we could use these machines for twelve hours per day, we would be able to test an additional 864 billion keys per day. That would account for roughly five percent of the daily total project activity in the first week of April. Adding twenty such labs would double the number of keys we could test per day. How many such labs were in the country would be anyone's guess. Thousands, maybe tens of thousands. If we could harness even a small percentage of these networks, we could test keys at a blindingly fast rate.

In early April, I spent a lot of time thinking about how to harness the power of systems that operated behind network firewalls. The basic architecture of clients communicating with the keyserver via UDP datagrams simply wasn't working—the firewalls were preventing those messages from flowing normally. We didn't really expect that we could convince most sites to make changes to the way that their firewalls worked to allow more free access between their internal systems and the public Internet. After all, those firewall systems had been put in place for a reason.

In the first years of the Internet's existence as a research project in the 1960s and 1970s, practically all Internet users knew each other. As the network grew in popularity among researchers, the community became less heavily interwoven, but there was still a high level of trust among Internet users in general, since all had a vested interest in making the Internet itself work. By 1997, though, those days were long gone: the Internet was no longer an academic or scientific curiosity. As the global network of networks became easier to use without highly specialized knowledge of arcane languages and computer technology, throngs of new users joined the Internet population. In time, the Internet population as a whole started to look less and less like the technical wizards who invented it and more like the general population, with its share of people with both good intentions and bad.

When communities are first settled, locks on the doors and windows are uncommon. As communities grow, along with inability to know and to trust people nearby, the use of locks increases. No one in New York City seriously considers keeping an apartment unlocked. Likewise, computer administrators lock up their networks (with firewalls) to deny access from unauthorized individuals.

A firewall is simply a barrier that separates one major unit (like the engine compartment of a car) from another (e.g., the passenger compartment). The idea of creating the separation is to avoid exposing the entire system to a threat affecting one unit or another.

When it comes to computer systems, firewalls on networks do the same thing: they separate one logical unit, or "zone" (e.g., the public Internet), from another (e.g., an internal network). Generally, this works by having a special computing and networking device have two network connections: one to the Internet and one to the internal network. With no other connection between the Internet and the internal network, any attempts to get traffic from one network to the other would need to pass through the firewall. Instead of allowing everything to pass—as happens with usual networking equipment—the firewall will look at the data to see if the policy programmed in by the administrator will allow the traffic to pass through.

With a few important exceptions, most firewalls would not allow UDP datagrams to pass. So, while our lab machines at universities and on networks directly connected to the Internet were able to talk to the keyserver, machines protected by firewalls would have their UDP-based messages to the keyserver blocked. The keyserver would never see the messages from the clients, simply as a matter of network policy forbidding traffic that the firewall administrators didn't know about from flowing between zones.

A lot of participants whose computers were behind firewalls were asking for details on how the DESCHALL clients talk to the keyserver, so they could approach their firewall administrators and ask for the configurations to be changed to allow DESCHALL traffic. Many users had made similar requests for other things, such as streaming audio, so finding at least a few sympathetic firewall administrators didn't seem out of the question. For participants who had administrative control over their organizations' firewalls, enabling DESCHALL traffic to pass was easy enough to do. Since I had authority over my company's firewalls, I was able to make the changes needed to allow a some of our machines to participate without opening the entire internal network to threats that the company wanted to avoid.

In most cases, though, the DESCHALL participants were not the same people who were responsible for their network firewalls. Going through the process of getting a change made—finding the right person, explaining what DESCHALL was about, and waiting for approval on a new policy—would take far longer than most participants would be willing to invest. In many cases, even if participants could make their cases to firewall administrators, their companies might decide not to participate as a matter of policy.

Firewall systems were becoming increasingly popular, and the number of networks where people could participate in a project that could not accommodate firewalls continued to shrink. We just had to find a way to make our systems work with firewalls and we knew that we had to find a way to do it without requiring special work by firewall administrators. If only we could find a way to make DESCHALL communications look like something that the firewalls already knew about and were configured to allow, we could get so many more clients running.

Thursday, April 3, 4:55 P.M.
Lawrence Livermore National Laboratory, California

As Karl Runge read his e-mail, he came across a message that Guy Albertelli from Ohio State had posted a few hours earlier. Albertelli looked through the DESCHALL project statistics for April 2 and found on the report a single system that tested over 700 billion keys and asked if anyone knew what kind of a system could test so many keys.

Runge did some quick calculations in his head and thought that it might be one of the latest Intel Pentium systems, with four processors running at 200 MHz each. Even that did not seem quite right, and he thought that it was more likely a system that was actually acting as a *proxy* or go-between for a group of systems that could not talk directly to the keyserver. Runge knew something about how individual project participants might introduce an architectural components like a proxy to overcome obstacles they faced in getting their systems to be able to participate in the DESCHALL project.

Five days earlier, Runge worked out a scheme to allow his computers at home to talk with the keyserver. His home local area network (LAN) had several systems on it and the entire LAN was connected to the Internet over a modem. That usually served Runge's needs very well, but sometimes UDP datagrams like the DESCHALL messages would not make it to their destination—UDP was at its most unreliable when trying to work over networks with severe bandwidth limitations, as could sometimes be the case when several computers were sharing the bandwidth of a single modem. Runge needed to find a reliable way to get DESCHALL messages from his computers running the clients on his home network to the network in his laboratory, where his modem connected.

Runge's solution was a pair of programs written in the Perl programming language. The first program would accept messages from the DESCHALL clients on his network in their usual UDP format and convert the message TCP format. (This would be like taking a message written on a postcard and putting it into an envelope.) The program would then forward the TCP message over Runge's modem link to another system that he had at work. That system was running another program—one that would take the message out of its TCP format and put it back into UDP, and forward it on to the real keyserver. Responses from the keyserver would go through the same process in reverse.

The system that actually handed the message to the keyserver would be the one that the keyserver thought was the client. Thus, both of Runge's machines would look like a single client to the keyserver. Runge thus concluded that whoever processed 700 billion keys the day before probably had a setup like his rather than a single top-of-the-line system with multiple processors doing nothing but running a DESCHALL client.

Unbeknownst to Runge, Justin Dolske and I had been privately talking about the same thing with Rocke Verser. Our system for working through firewalls was a straightforward one, almost identical to the one that Karl Runge created for his own system. Justin Dolske wrote a pair of programs: one was called U2T (UDP to TCP). Instead of converting a UDP datagram into a TCP packet, U2T formatted the data to look like a Web request—a HyperText Transfer Protocol (HTTP) message carried inside of a TCP packet. The other was called T2U and converted the TCP-based HTTP message back into a UDP datagram to be forwarded to the keyserver.

The Web uses HTTP to communicate. HTTP is a higher-level protocol than TCP and UDP: it relies on foundation provided by "lower-level" protocols. Protocols like TCP and UDP will get the data you need sent from one system to another in the appropriate chunks, actually carried across the Internet infrastructure inside of IP packets. HTTP defines the format of the message itself. It would be the equivalent of the sender and recipient deciding that when they are sending messages back and forth, they're always going to have some lines at the top like `To`, `From`, and `Date`. They would also need to agree to write in the same language. This is the role of HTTP.

So in practice, what happens is that Web traffic is formatted in HTTP, carried in TCP packets, which are in turn carried in IP packets.

Participants who wanted to run clients for DESCHALL behind their corporate firewalls would download U2T and run it on a machine on the same internal network as the clients—behind the firewall. The user would then tell the U2T server the address of the firewall system used to forward Web requests from internal systems to external (i.e., Internet) Web sites. Once the U2T system was configured and started, it would start listening for UDP datagrams that had DESCHALL client requests in them.

After the U2T system was running, the users would then start their DESCHALL clients, but instead of telling the clients to use the real DESCHALL keyserver—to which the firewall would block access—the user would tell the DESCHALL client that the U2T server was the keyserver. That would start the client, which would contact the U2T server and ask for keys to test.

The U2T server would receive the message in a UDP datagram from client and put exactly the same message in the form of an HTTP message, which would get put into a TCP packet, which would get put into an IP packet, and then sent to the firewall with an ultimate destination of one of the three T2U servers that Justin Dolske and I ran at Ohio State and Megasoft, respectively. Each T2U server had the same functionality as the others; we just used three servers so we could spread the load among more systems.

The T2U server would listen for HTTP messages inside of TCP packets from a T2U server. Just as the participant's U2T server took the client request out of the UDP datagram and put it into an HTTP message, the T2U server would take the client request out of the HTTP message, create a new UDP datagram, put the client request into that datagram, and then send that datagram to the keyserver.

Next, the keyserver would receive the client request in a UDP datagram that looked like any other client request. It would accept the request, and send the result back to the T2U server that sent the request—in the form of a usual DESCHALL message in a UDP datagram.

The T2U server would accept the response from the keyserver, pulling the message out of the UDP datagram, creating a new HTTP message and putting the response from the keyserver into that HTTP message. That HTTP message would then be sent back to the U2T server, which would then pull the response out of the HTTP message, put it back into a new UDP datagram which would then be sent to the original client.

Adding the T2U servers to the architecture and distributing U2T software that participants would run behind their firewalls satisfied all of our requirements: DESCHALL could work safely through firewalls without requiring any changes to be made in the firewalls, the DESCHALL clients, or the keyserver.

Having addressed the issue of how to help clients behind firewalls to participate, we were ready to charge ahead. With quadrillions of keys left to test, we were going to need the help of people behind corporate firewalls.

15

Progress

Wednesday, April 9, 8:09 P.M.
Yale University, New Haven, Connecticut

Computer science student Jensen Harris had two machines with some extra processing power available. Like most student-owned machines, they weren't spectacularly powerful. Both had Pentium processors, one 90 MHz and the other 150 MHz—but they spent most of their time sitting idle, so Harris thought it would be a good idea to contribute their idle cycles to the DESCHALL effort.

Computer microprocessors come in many varieties, such as Intel's Pentium, the PowerPC from IBM and Motorola, and Sun Microsystems' SPARC family of processors. While the details on what exactly happens inside vary dramatically from one family of processors to another, all processors essentially work the same way: an instruction is given and the processor responds by performing calculations or moving data from one place to another. Processors have a "cycle"—a tiny period of time during which an instruction can be executed.

"Hertz" is the metric measurement of frequency, named in honor of German physicist Heinrich Rudolf Hertz, who made several important contributions in the field of electromagnetism. One hertz (1 Hz) is one cycle (or event) per second. One kilohertz (1 kHz) is one thousand cycles per second. One megahertz (1 MHz) is one million cycles per second. One gigahertz (1 GHz) is one billion cycles per second.

Processor clock speed can be a useful measurement to compare processors of the same family to each other—a 150 MHz Pentium is about 60 percent faster than a 90 MHz Pentium. The problem with clock speeds is that they are almost useless for comparing processors of different types to each other; there is no way to tell how a 100 MHz PowerPC processor compares to a 100 MHz Pentium, since what each processor will be able to do in a given cycle will vary dramatically. Some processors, like the PowerPC will do a lot of work in a cycle, while others like the Alpha will do very little in a single cycle.

Watching the DESCHALL client run was a great way to get a sense of how many different things contribute to just how much work different computers can accomplish in a given period of time. Because the DESCHALL client would simply test one key right after another without waiting for input from the user or anything else, the client would just run and run as fast as the processor could support it. The processor's clock speed will matter to overall system speed, but so will the amount of work that can be done in a given cycle. How much can be done in a cycle depends on things like how the processor is designed and just how well the software can take advantage of that design. Rocke Verser's hand-optimized DES key testing software for the Pentium processor was so fast because he was able to get more work out of each processor cycle.

Looking over the project statistics for the past few days, reproduced in Table 4 (on page 89), Jensen Harris saw how powerful DESCHALL had become. At this point, we were testing 2 trillion keys per day, by comparison to the 496 billion keys we were testing per day less than six weeks earlier. Harris wondered about the value of having his two mid-range desktop computers working on the project. It was clear that there were many of other locations doing a lot more work. But at what point, Harris asked in a message written to the DESCHALL mailing list, does a contribution become too small to be worth the effort?

DESCHALL participants answered Harris' question resoundingly: every little bit helped. Unless the keyserver simply could not keep up with demand for instructions from key-testing clients, even the slowest of machines was valuable.

With Rocke Verser's fast Pentium software, however, the danger of any Pentium machine ever becoming a burden was nonexistent—the DESCHALL key testing software ran much more slowly on many other systems that used other processors. Those slower clients would become a burden on the keyserver long before even the slowest of Pentium processors. With the lightweight UDP-based protocol for communication between clients and servers, the likelihood that the keyserver wouldn't be able to support the load put on it by the number of clients we had was also pretty low.

Lee Sonko from Bowne Global Solutions was in agreement with the rest of the mailing list participants, but he wanted to see just *how much* the small contributors mattered to the project.

Using the statistics from April 8, Sonko was able to show how much of the day's work was performed by, as he called them, "big boys," "medium-sized domains," and "small domains." The "big boys" were the five domains testing a trillion or more keys per day. The "medium-sized domains" were the 55 domains testing between 69.8 billion and 1 trillion keys per day. The "small domains" were the rest—108 domains testing between 16 million and 66.8 billion keys for the day. After breaking them up into those three groups, he totaled the number of keys tested, not per domain, but per group. (Table 5 summarizes his findings.)

Group	*Keys Processed*
Big Boys	11.16 Trillion
Medium	10.95 Trillion
Small	1.83 Trillion
Total	23.95 Trillion

Table 5. Work Performed by Size

Although the small domains were testing only one tenth of the number the large domains were testing, it was clear that their contributions mattered. None of the DESCHALL participants wanted to lose any processing power at all.

A total of almost 24 trillion keys were processed on April 8 by all DESCHALL participants together. In the same day, roughly 1326 machines (as determined by unique IP addresses, a reasonable but inexact approximation) worked on the project. That would mean that on average, each machine processed roughly 18 billion keys that day.

We used a relatively modest system as our "benchmark" to measure how fast a "typical" machine would work its way through the DES keyspace. That benchmark system was a 90 MHz Pentium running FreeBSD, which ran at a rate of 454,000 keys per second. In a 24-hour day, that machine would process some 39 billion keys. Thus, a relatively modest 90 MHz Pentium computer, working all day, would be faster than average, meaning that it would increase the average speed per host.

We were nowhere near running at the level of that any client would prevent any other from getting work done. Not only were we not ready to ask people with the slower systems to stop participating, we wanted more clients, and we needed as many as we could get. Testing 72 quadrillion keys was a job for a lot of processors—the more, the better. As Justin Dolske observed, "Quantity has a quality all its own."

While we were also working to support participation through firewalls, other projects were still searching for the key. The European SolNET effort was still going strong. The DES Violation Group was also still running, though by this point they were falling further behind. Each of these efforts were, in one sense, competing with our effort, since they ran their own keyservers and their client software was different from ours. If DES Violation Group managed to find the key, they would get the prize money that RSA put on the table. Even so, RSA's DES Challenge was a bit like a scavenger hunt where after there was a winner, a prize would be shared with everyone: any team finding the right key would hasten the development of a stronger standard to replace DES—a prize for everyone—but we were all hoping to be the one that found the key.

Interestingly, DES Violation Group, like DESCHALL, was based in North America and made the same decision that we did—to restrict the distribution of clients to the U.S. and Canada. Also like DESCHALL, DES Violation Group used the North American Cryptography Archive to distribute its client software without violating U.S. cryptographic export policy.

Of the U.S.-based systems for participating in the DES Challenge, DESCHALL was most effective in its publicity, because, among other factors, the tremendous speed of our clients and our participants' zeal in recruiting their colleagues and friends to join the project. Unfortunately, the fact that there were clients in the archive for DES Violation Group (with names beginning with "des") and clients for DESCHALL (starting with "deschall") was confusing. That confusion led some people who heard about DESCHALL to go to the site to download the client and pick up the first DES Challenge client in the alphabetical list—which was for the DES Violation Group.

Friday, April 11, 1:56 A.M.
Rensselaer Polytechnic Institute, Troy, New York

RPI student Bill Moller understood why groups in the U.S. were not trying to get people from outside of the country to help on their efforts. He could also see why cryptographers in the U.S. and Canada were not especially interested in just sitting back and waiting for a group from Europe or somewhere else to prove the point that they have been trying to make for years—that 56-bit keys are just too small and are subject to brute-force attacks. What Moller did not understand is why there were multiple U.S.-based projects working on the same goal. Presumably, the people working on both of these projects were under no legal restriction that would prevent them from working together.

Moller posted his concerns to the DESCHALL mailing list. In particular, he wanted to know just why two U.S.-based groups were competing instead of working with each other. Apparently he figured that, say, 1000 users working together on one coordinated effort would find the key faster than 1000 users spread across two or three competing projects.

Answers to Moller's question proved interesting. Some pointed out that only DESCHALL supported their platform. (This was something we heard over and over again throughout the course of the project, particularly from the users of operating systems like OS/2. Although relatively small in number, these users were advocates, able to draw large numbers of others to the project.)

Some participants pointed out that having multiple groups working on the same problem was good insurance, so that if a group had a problem with its software, or for some other reason failed to find the correct key, the other groups could pick up the slack. Still others highlighted the possibility of rogue clients falsely reporting that they had tested certain keys, so having one project test keys that other projects had presumably tested wasn't always necessarily a bad idea. In fact, Peter Trei's uncoordinated DESKR software was developed specifically to deal with these kinds of problems.

Rocke Verser suggested that if competition were a problem (and he didn't say that it was), the question of why we were "competing" should be directed to DES Violation Group, rather than DESCHALL. The DESCHALL keyserver had been online since January and clients were available to the public since February. We didn't really know where DES Violation Group came from or the people behind it, but it seemed to emerge some time after DESCHALL was already up and running.

Despite the confusion about clients from two groups being distributed from the same archive, DESCHALL continued to widen its lead over DES Violation Group. The European DES-Challenge group that spun off from Germano Caronni's 48-bit RC5 crack had been written off by even most of its organizers, who had since joined the SolNET effort.

DESCHALL was the fastest DES Challenge contestant, and was getting faster almost every day. The basic statistical information we did report—number of keys tested per day and total number of keys tested to date—was enough to show our lead, but many participants wanted to see more details. Much of this desire was probably inspired by some competing projects, which really did have extensive statistical reporting mechanisms on their project Web sites, showing breakdowns per domain, creating line graphs of various participants' effort. DES Violation Group in particular had extremely impressive graphical reports that looked much like the charts generated to support analysis of stocks and bonds. Oregon State's Adam Haberlach quipped that DES Violation Group was probably using more CPU time for computing stats than they did for testing keys.

Although DESCHALL had by far the most primitive statistical reports of any coordinated effort, we were taking the need for better reporting seriously. Justin Dolske created a program on April 4 that would take the previous night's raw data—as posted on Verser's project status Web page—and create graphs that would show our progress, showing keys tested per day, and total keys tested. These graphs charted the progress of the project as a whole. Participants liked to see how the project was doing overall, but they wanted more. They really wanted to see how their organizations were doing individually.

Dolske continued to work on ways to present DESCHALL statistics graphically, hoping to find a way to let participants see their own contributions over time. On April 21, Dolske announced his work and that he was looking for people to test his dynamic graph-generating software that came to be known as "Graph-O-Matic." Rather than seeing only the progress of the entire project, Graph-O-Matic was a web-based utility which provided data for any domain whose progress you wanted to check, and receive back graphs showing the progress of those domains only.

The day before Graph-O-Matic opened for testing, Karl Runge from Lawrence Livermore National Laboratory posted a note of his own to the DESCHALL mailing list about some of the work that he had been doing in looking at the project statistics. He had been using the reported statistics to generate some graphs that could be used for analysis that would help us understand what to expect in the days and weeks ahead. By mid-April, our key testing rate really began to pick up—doubling every seven to eight days for the past month. Runge showed our progress on a graph plotted atop a mathematically proper exponential curve showing that participants who called our growth exponential were not exaggerating.

Using the graphs to show past progress and to project future rates, we could see that despite the tremendous amount of work that remained, there was light at the end of the tunnel. In fact, if we could sustain our exponential growth, we would have have half of the keys tested after eighty-seven days of work (only fifteen days away!), and the entire keyspace tested in ninety-six days, about the end of the first week of June.

Runge himself called attention to a problem with using his projections to show when we would finish testing half or all of the keyspace. On April 18, we just finished the milestone of having tested one percent of the total DES keyspace. As any statistician will confirm, taking performance for the first one percent of a project like this and using it to predict the next 49 (or 99!) percent would be stretching the data. The model for analysis was fine—but we were just too early in the project to have enough data to show things like how the project would react if it ran into any trouble.

After a month of exponential growth, we were starting to get things moving. We needed to sustain the growth in order to continue pulling ahead of the competition.

16

Trouble

***Saturday, April 12, 11:42* P.M.**
Rutgers University, Piscataway, New Jersey

At Rutgers, like universities all over the country, the semester was starting to wind down. While the year had not officially drawn to a close, in the minds of students and probably more than a few of their professors, it was effectively finished. Student Scott McIntyre started thinking about summer vacation—going back home, visiting old friends, and spending time with family. As he thought about what summer would bring, he realized that many students who were participating in DESCHALL might not be able to be involved in the project over the summer, particularly if they didn't have ready access to the Internet or large numbers of more powerful computers.

McIntyre decided to post his observation that, in another month or so, students were going to be heading home for the summer, and predicted a likely decrease in DESCHALL's performance. The message set the mailing list abuzz and the group quickly started to look at our top key testers. (The top thirteen key testers for the day are shown in Table 6.) Ten of the top thirteen contributors were universities and another was a high school. If all of the students running clients on those sites were leaving for the summer, they would take a lot of our computing power with them.

Mailing list participants debated the severity of the problem. Some, including Drew Hamilton of Strategy Management Laboratory Corp. reasoned that it wouldn't make any difference, since students were going to be running the software on their own systems, and it didn't make

Keys tested	*Clients*	*Contributor*
4.26 Trillion	275	Oregon State University, Corvallis
3.31 Trillion	117	Rensselaer Polytechnic Institute, Troy, New York
1.98 Trillion	1	DESCHALL U2T/T2U Proxy Users (Aggregate)
1.64 Trillion	91	Michigan Technological University, Houghton
1.48 Trillion	197	Ohio State University, Columbus
1.36 Trillion	61	Rochester Institute of Technology, New York
1.17 Trillion	164	Brigham Young University, Provo, Utah
1.06 Trillion	39	Worcester Polytechnic Institute, Massachusetts
0.77 Trillion	14	Apsylog Development, Nanterre, France
0.77 Trillion	27	Duke University, Durham, North Carolina
0.72 Trillion	27	Michigan State University, East Lansing
0.71 Trillion	11	University of California, Davis
0.71 Trillion	28	Thomas Jefferson High School for Science and Technology, Alexandria, Virginia

Table 6. Top Thirteen Key Testers for April 12

any difference whether those systems were plugged into the university network or not.

Others, including Brian Osman, a student helping to coordinate DESCHALL participation at RPI (Rensselaer Polytechnic Institute), pointed out how much of the total key testing was being done on student-owned machines that would almost certainly need to use inconvenient dial-up connections and compete for time on the telephone line with other family members. Without easy Internet access, many of those systems would probably not be running the clients that they would if they were sitting in the dorms wired for constant high-speed access delivered through permanent local area networks.

Being among the optimists, I suggested several things that would help us to be able to continue to test keys at a phenomenal rate. First, we needed to develop optimized clients for architectures other than those based on Intel's 80486, Pentium, and Pentium Pro processors. Those kinds of systems from providers like Sun Microsystems, Silicon Graphics, Inc. (SGI), and Digital Equipment Corporation (known generally as DEC) were plentiful in research and academic laboratories. Even with the comparatively few graduate and summer students who would still be using them, university computer labs would be doing much less "real work" than they did the rest of the year, leaving more processing power available for testing keys. Getting client software more heavily optimized for these systems, many of which were already running DESCHALL clients, could greatly improve their efficiency. If those systems' clients could be made to be as efficient as the Intel clients—which would mean performing the optimization work manually—they would be dramatically faster than the Intel systems.

Second, we could raise awareness among other computer users who would not only run the client but would also get others to run the client. Internet Service Providers (ISPs) might be a good way to get many others interested. Many ISPs created little online communities for their users, and if we could convince ISPs to encourage their users to run the DESCHALL clients, we might be able to fan a few flames of friendly rivalry among local and regional ISPs that we managed to stoke among universities. Perhaps ISP system administrators would take to checking the DESCHALL project statistics day after day, trying to get their networks higher in the rankings. We had not seriously attempted to recruit ISPs up to this point, but the idea seemed reasonable enough and in any case was worth considering as a source of potential computing power.

Finally, once DESCHALL gained the ability to work nicely with firewalls, corporate environments, and others whose systems were previously unable to participate would be able to join in the effort. Justin Dolske had written the code at this point. Rocke Verser and I were testing the software and working with Dolske to make the proxies ready for production. We were less than a week away from making the firewall-friendly proxies available. The way I saw it, we had plenty of opportunities to get new participants running the clients, so even if we did lose the academic contribution, we could keep growing.

Dolske didn't think that we needed to worry about losing the universities over the summer. Since he was a grad student and would not be leaving for the summer, he would continue to run the client at Ohio State. He suggested that school contributors simply find grad students who would be around all summer and recruit them to keep the clients running.

Besides, the statistical analysis Karl Runge just performed showed that if we could maintain our exponential growth over the next fifteen to twenty days, we'd almost certainly find the key before summer got underway. The summer vacation problem could prove to be, well, academic.

Wednesday, April 16
Yale University, New Haven, Connecticut

"Well, these are the newest computers we have and we don't want to wear out the processors."

Computer science student Jensen Harris was glaring in disbelief at the computer lab manager. With the blessing of the lab's manager, Harris had been running the DESCHALL client on fifteen Linux workstations for the past day. Each of these systems was an HP with a 166 MHz Pentium processor. After watching how the process ran, the lab managers decided to stop them and to tell Harris that he was not allowed to use the systems for DESCHALL. The systems sat completely idle, all day long, waiting for the beginning of a project that would not start until June.

"It probably also voids our warranty with HP to run programs like this because it is an undue strain on the processor."

Harris could not believe his ears: he knew perfectly well that whether the computer was sitting idle or performing the world's biggest computation was irrelevant to the processor. For each cycle, the processor would simply perform an operation. If it had no operation at all to perform, it would perform a "no-op," an operation that means "no operation" and simply keeps the processor occupied until the next cycle starts.

The lab manager continued, "Processors like these are only 'rated' a couple of thousand cycles per minute—going over that is not something we're about to attempt without studying the effects beforehand."

Wisely, Harris set aside the urge to engage in an act of violence. A processor of 166 MHz was indeed 'rated'—for 166 million cycles per second—and nothing that could be done in software could change that. DESCHALL had some amazing software, but the clients weren't magic; our client just couldn't make a processor run at a higher speed.

"Tell them that if they can make a nice client that doesn't run the processor too hard, we'd be happy to help."

Harris knew that not much could be done with someone who was both nice enough to look at the problem and to offer help if some accommodations could be made but clueless enough to think that the processors could 'wear out.' He walked away, shaking his head. At least he could run the client on the computer in his dorm room.

Friday, April 18, 6:13 p.m.
Northwestern University, Evanston, Illinois

Yale was not be the only university to ban DESCHALL clients. Computer science student Vijay S. Gadad reported that his department at NWU wasn't interested in working on the DESCHALL project, as they couldn't see "why they should be helping RSA."

Apparently, some administrators were under the impression that this project was somehow helpful to a private, for-profit company, and that their resources should not be used. Upon learning that RSA was actually giving prize money, administrators further objected on the grounds that students would be participating in the effort for personal profit. Administrators would often not listen to the explanation that the university itself would actually get the money.

Many DESCHALL participants reported that they also had difficulty getting system administrators and operational staff to understand that the software was actually quite friendly to the system. Many would just see the system devoting a lot of time to a single program and assume that it was somehow hurting something. But nothing topped Harris' experience with the staff who thought that an idle CPU cycle is somehow less demanding than a "busy" one.

Some participants were quick to heap scorn upon the lab managers who obviously had no understanding of how computers worked. For me, these stories were an eye-opening experience—I assumed that everyone who worked with computers every day simply knew how they worked. The DESCHALL project was providing an excellent opportunity for participants who didn't know how computers worked to learn. But more importantly, we had DES keys to test, and we needed to get to get more people involved. We could allow ourselves to become downhearted if we only looked ahead at how much work yet remained. Watching a few milestones go by would help us to stay motivated.

17

Milestones

Thursday, April 17
Loveland, Colorado

DESCHALL reached a significant milestone when we finished testing the first one percent of the keyspace. Seven hundred twenty-one trillion down, a few quadrillion to go.

To draw attention to the achievement, we issued our second press release. After quickly describing the project for readers who had not yet learned about DESCHALL, we wrote:

> According to Rocke Verser, a contract programmer and consultant, who developed the specialized software in his spare time, "There are over 2500 computers now working cooperatively on the challenge."
>
> Using a technique called "brute-force," computers participating in the challenge are simply trying every possible key. "There are over 72 quadrillion keys. A number," Verser quips, "about 15,000 times larger than the deficit."
>
> But the DESCHALL group is racing through the keys at an incredible pace. The group is now trying over 50 trillion keys per day—or more than 600 million keys per second.
>
> Perhaps even more impressive, the number of computers participating, and the rate at which they are trying keys has been doubling every eight to eleven days for the past two months.
>
> If the number of participants continues to double every ten days, it should take about two months to find the key. If no

> other participants joined the effort, it should take about two years to find the key.
>
> Word of this cooperative effort has spread primarily by word of mouth and the Internet equivalents—IRC, newsgroups, and mailing lists.
>
> No one knows where the growth of this type of cooperative computing effort will peak.
>
> "Members of the DESCHALL team will be in a festive mood, Friday," Verser predicts. "About supper time" on Friday [April 18], DESCHALL computers will have tested one percent of the total set of 72 quadrillion keys.

As our machines continued to test keys, we put the notice up on the Web site and participants began to distribute it to their local media outlets. We passed the milestones as expected.

Monday, April 21, 9:39 A.M.
Megasoft Online, Columbus, Ohio

One of my DESCHALL clients finished testing a block of keys and sent a message to the keyserver that it was finished and was ready to receive another block. After a few minutes went by, the client attempted to send the message again, still without response. Working on something else, I did not immediately notice that the client was having trouble. About an hour after the client started waiting for more work to do, I returned to my e-mail.

Several participants sent messages to the DESCHALL mailing list to see whether others were having difficulty reaching the keyserver. I checked my clients and saw one waiting. I quickly started a network utility called traceroute, which would show response times between my system and every hop along the way to the keyserver. With traceroute output, I would be able to tell whether my systems could reach the keyserver, or if not, just how close I could get.

The results showed that not only was the keyserver unavailable, but a whole section of a network that connected the keyserver to the Internet were offline. I picked up the phone and dialed Rocke Verser's home number.

Monday, April 21, 7:40 A.M.
Loveland, Colorado

After a long night of developing new DESCHALL clients, compiling project statistics, managing server logs, and coordinating the efforts of other client developers, Verser was sound asleep when the phone rang.

He picked up the phone to hear me announce that the keyserver and its network was offline. After thanking me for calling, he went to the keyserver in his home office and saw something that he did not frequently see: the modem connecting his home office network to the Internet was offline and not correcting itself. Although using a modem, Verser did not use it as standard dial-up customers did; he had dedicated service. The modem never disconnected unless there was a problem, and if that happened, his system would recognize it and immediately reconnect.

After some quick work, Verser restored the connection, and the keyserver became visible to the Internet. Messages, including the one from my client that had been waiting, started to flow in once again. The outage, which lasted for approximately three hours, was long enough for only a few participants to notice. Certainly the problem could have been much, much worse.

Verser watched to ensure the system was working properly, typed an e-mail message to project coordinators describing the problem and his assessment of its impact, and went back to bed.

Less than one week after passing our first major milestone (completion of the first one percent of the keyspace), we passed another: we finished testing 1 quadrillion keys on Tuesday, April 22. But for the remainder of that week, our growth rate began to slow. Although still growing at a fast pace, graphs of our progress started to show that the exponential curve wasn't holding.

Michael J. Gebis from Purdue looked at the DESCHALL "keys tested per day" statistics on April 24 and noted that the number of keys tested was starting to resemble a sigmoid curve. In a sigmoid (basically S-shaped curve), the first part looks exponential, but then the curve becomes linear until it reaches a peak and then it levels out. He suggested that it was possible that we had passed our exponential growth spurt, and we might see only linear growth in the future. Although the prediction was disappointing considering the progress made over the previous weekend, no one looking at the data could seriously advance a more optimistic argument.

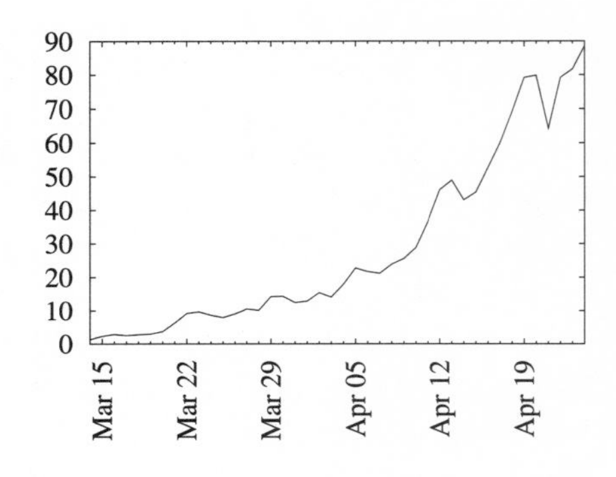

Fig. 5. DESCHALL Keys Tested Per Day, March 14–April 24 (in Trillions)

Thursday, April 24, 3:57 P.M.
Lawrence Livermore National Laboratory, California

Concern over the reduction in our growth rate led some participants on the mailing list to raise questions about competing projects "wasting" time and effort. Karl Runge, our resident statistician, started crunching some numbers to get some sense of how long it would take to find the right key with two DES Challenge projects. He compared the average search time for two cooperating servers to the average search time for two non-cooperating servers.

DESCHALL's overall computing power had grown tremendously in the past six weeks. From March 14 to April 23, our growth rate followed an exponential curve, even if the developments of the past few days made it appear to be unsustainable. At the beginning of that curve, we estimated that it would take us roughly seventy-five years to find DES keys on average, and by April 24, the figure had dropped to under fourteen months. Runge found that the average search time would be sixty-seven days (from March 14) for two cooperating servers. For two independent servers, average search time would be only about two days longer. This had a lot to do with the number of keys being tested toward the end of the project, which was the highest part of the end of the exponential curve.

Having two cooperating projects would make more sense in the case of linear growth, since it would spread the key testing out over a larger period of time. With cooperating keyservers following the linear growth that was apparent from April 11 to April 23, the time to find the key would be 109 days (from March 14) on average. Non-cooperating servers starting on March 14 would find the right key 121 days later on average.

The diference was not very large. Obviously, a prolonged effort with a lower growth rate over time would benefit from using multiple, cooperating keyservers. However in our case it was clear that it didn't really make sense for us to join forces with another group. Runge observed that after all the math was done, he had learned what his two children had been telling him for years: it doesn't pay to cooperate.

We didn't have to wait long for Runge's words to prove themselves. Within a few days, DESCHALL went on to set many new records. On Friday, April 25, the project maintained a key testing rate of over 1 billion per second for an entire 24-hour period. On Saturday, April 26, we tested more than 100 trillion keys in a day for the first time. At that rate, our key testing rate was such that if simply sustained, we would reach the halfway mark in under one year.

By Sunday, April 27, we had tested two percent of the entire keyspace. Testing the first one percent of the keyspace took us ninety days by RSA's count. Testing the second percent had taken a mere ten days.

By the end of April, DESCHALL had transformed from a single person in Colorado with some fast DES key-testing software to a fully-functional virtual organization, with development, project management, communications, and recruitment functions being addressed by volunteers. Still more people started to take note and we began hearing from people who could not participate but wanted to encourage us, or whose participation would be limited.

The most visible gap in our client offerings was for Macintosh machines. We had been getting a great deal of help from the enthusiastic OS/2 crowd. Especially since we began to report statistics showing how many keys were being tested per platform. We were pretty sure that we could get a big boost from the availability of a Macintosh client, who were also known to advocate their favorite computing platform with zeal. Many writers in the mainstream computing media had written off the Mac and OS/2 as irrelevant, figuring that "everyone" used Windows. The users of these marginalized systems often sought opportunities to contradict the myth of Windows ubiquity, and any study or project that would rank participation by platform provided the kind of data that the enthusiasts needed to show that their platforms were a force to be reckoned with.

To find how much interest there was in a DESCHALL client for the Macintosh, pre-cognitive science undergraduate student TC Lai at the University of California at Los Angeles posted to the DESCHALL mailing list. After getting some favorable responses and signing the necessary confidentiality agreements protecting the DESCHALL source code with Rocke Verser, Lai started work for a DESCHALL client for the Macintosh that was first released on April 29.

Mac users weren't the only ones clamoring for new clients. Users with relatively uncommon architectures (such as Solaris/$x86$, Sun's Unix implementation designed to run on Intel-based systems instead of its own systems with the SPARC processor) also had lots of processing power available and wanted to join in, as did users of various scientific and engineering workstations. Almost all of those requests would be answered in time.

People who were watching our progress closely were interested in an important number that we simply did not have—the total number of computer systems that were contributing processing power. We did not have this figure because the key-testing clients did not have a serial number. What we could do, however, was approximate, based on several figures that were available to us. The keyserver was keeping track of the unique Internet Protocol (IP) addresses and counting how many addresses were checking in; we kept track of this information and reported it in the daily statistics. However, these reports did not indicate how many processors were testing keys on multiprocessor systems. Also, all clients behind proxy-style firewalls were reported as a single client, since all of their data would arrive at the keyserver from the firewall, rather than from the client running the software.

In addition, participants who were using dial-up Internet access with dynamic IP address assignment found that if they connected to the Internet, say, ten times per day, they might well have ten different IP addresses. Dynamic addressing was typical for home users; since their systems were not advertising the availability of any information to the Internet population, there was no need for its address to remain the same for any significant period of time. Static addressing was used for dedicated lines, whether they ran over high-speed lines leased from the phone company or whether they ran over low-bandwidth modem lines that were permanently connected, like Rocke Verser's home office that housed the DESCHALL key server.

Some participants wanted to count the number of clients instead of using these estimates. One method to get such a count was to put a serial number into each client, so the total could be counted accurately. Having clients report their serial number when checking into the keyserver would require a change in the communication between key-testing clients and the keyserver. Since our overriding concerns were focused on cracking the DES challenge message, protocol changes were a last resort and would only be employed when addressing a problem important to the project's core function or stability.

In the end, we just reported the numbers that we had, unique IP addresses, and pointed out that some forces (like firewalls) tended to deflate the number of clients, whereas others (like dynamic IP addressing schemes) tended to inflate the number.

The matter of reporting individual contributions to the project came up again, although Graph-O-Matic had solved the biggest concern of this type some three weeks earlier. Since Graph-O-Matic generated its reports from the DESCHALL statistics, it reported contributions of keys tested by domain name. Some participants wanted to be able to organize themselves into teams or to have their contributions from several different networks (such as home and work) be counted together. SolNET provided this service by having its client software report the e-mail address of the client operator along with the results of its testing. Users could then configure all of the clients they wanted to be counted together to use the same e-mail address, and enter their e-mail address into SolNET's graphing server. Instead of getting just one network or another this way, all of a user's clients statistics would be aggregated together in the report. Since DESCHALL didn't send e-mail addresses between clients and servers, Graph-O-Matic could not provide that functionality. Although DESCHALL coordinators agreed the feature would be nice, it would require a change in the protocol between the client and server—something that would pose too great a risk for us to try unless it directly solved a serious obstacle to the core problem, the testing of keys.

Another request that came in from DESCHALL participants was the ability to request larger blocks from the keyserver than the ones it was handing out. Dial-up users were accustomed to working offline and going through the hassle of connecting only when interactive use of the Internet was necessary. With the DESCHALL client running dial-up users had to reconnect to the Internet ever few hours (around the clock) or to let the machine sit idle after finishing a block while it waited for the next connection. Users who did not want to have their systems connect every other hour started to ask us to give their systems enough keys to keep them busy for longer periods of time—six, eight, or twelve hours. Again, this would require a change in the DESCHALL protocol so we did not implement the feature.

In reality, only individual machines were using dial-up modems and by this point, the only systems that could not automatically connect when the DESCHALL client needed to communicate with the keyserver were Windows machines. Windows users would simply need to wait for a while before their dial-up woes would be solved.

Finally, another type of request that DESCHALL participants repeatedly made had to do with optimizations. Highly sophisticated processors (even 64-bit processors) performed poorly by comparison to the hand-optimized 32-bit 486, Pentium, and Pentium Pro clients that we had. More correctly stated, the machines with the sophisticated processors were fast, but Verser's optimized Intel code was awesome—with half of the power of some sophisticated workstations, personal computers were able to get better than twice the performance. The requests that began in early April for optimized clients for the other architectures continued unabated through April. There was no doubt that we needed optimized clients for the 64-bit processors like Alpha, UltraSPARC, and MIPS, but we needed more help before we could deliver fast code for those users.

The second optimization-oriented requests came from users of the Intel "clone" processors from AMD and Cyrix. Ultimately, such optimization would require someone who understood the internal workings of those processors. No one with the expertise would ever step forward to do the work—there just aren't very many people who have that depth of expertise, especially who also had an understanding of how to implement a cryptographic algorithm like DES. It might even be true that at the time, a large portion of the people who had experience with both cryptography and such low-level code optimization were already at work on DESCHALL or projects like it. Consequently, the users of the so-called Intel-compatible processors would continue to use the Intel 486 client, because it ran the fastest on those systems—though still nowhere near the speed on the genuine Intel hardware.

18

Gateways

Sunday, April 20
Columbus, Ohio

Justin Dolske and I put the "T2U" (TCP to UDP) gateways into production and released the corresponding "U2T" (UDP to TCP) proxy software that would allow people to participate in DESCHALL by running clients from behind firewalls. Then we ran into trouble.

Dolske developed the U2T gateway software in the Perl programming language, which made it able to run on pretty much any kind of computer. This was an important requirement for gateway software, since participants on corporate networks could have any kind of systems in use and we could not put the same kind of effort into building and maintaining gateway software that we put into the key-testing clients. Perl software could be written once and work without modification on dozens of types of computers. One particular problem arose on Windows-based servers, however. Although Windows systems could run most Perl code, a critical feature that the U2T gateway needed was not present in Windows. In corporate environments where a Unix system could be used, they could just run U2T there. Users in Windows-only environments were stuck so we started to look for alternatives.

I was initially very supportive of the idea of having Windows-based gateways. After all, the job that we were trying to accomplish was huge and we needed as many clients as possible. If Justin Dolske's gateways wouldn't work for every possible participant's computing infrastructure, it seemed to make perfect sense that other types of gateways could be useful.

After seeing that Dolske's U2T software wouldn't work on his Windows servers, Bret Stastny started to work on U2T software specifically for Windows systems. On April 22, Stastny posted a message to the DESCHALL mailing list that he wrote a U2T gateway for Windows.

Since the Dolske gateway code was already released in the form of source code (the human-readable version of a program), thus showing very clearly how the internal gateway needed to communicate with the external gateways that Dolske and I were running, I assumed that other contributions of gateways would simply be replacements for the internal (U2T) proxy, making use of the T2U servers that Dolske and I had in production for the whole project.

Sadly, Stastny did not just write a U2T gateway. Instead, he wrote a completely new U2T/T2U gateway pair, using a protocol completely incompatible with Dolske's. Essentially, we would need to duplicate all of Dolske's proxy architecture on Windows to use Stastny's software. None of the DESCHALL coordinators had Windows servers in production, and given the instability and management difficulty of Windows, we didn't want to trust a critical piece of the entire DESCHALL project architecture to a Windows system. The production T2U proxies would have to be stable, largely self-managing systems. We could not take on the additional administrative load of an incompatible T2U proxy set. We asked Stastny to get his internal gateway to work against our T2U gateway, but that work ultimately did not lead to any success.

Other attempts were made at creating internal gateways for people who could not run Dolske's Perl-based U2T. On May 8, Aaron Williams at Adaptec posted a Java version of the U2T gateway. Java, like Perl, would run unmodified on dozens of computer types. Java did a much better job of hiding the underlying computers' differences better than Perl, and a Java gateway would work unmodified on both Windows and Unix systems. While the approach—a plug-in replacement for the internal gateway that used the production external gateways—was correct, there were some problems. Williams hadn't worked with us to coordinate the effort and to test the software, so as released, the software would send malformed updates to the T2U gateways. The gateways, in turn, forwarded the request along to the keyserver, which would ignore the message. Any work done by clients using the Java gateways would be lost.

The gateway problems didn't end there. Williams' subsequent attempts to get his gateway tested and debugged for re-release were also done without our cooperation. He was testing his software by having his nascent gateway send messages to the production T2U gateways, which dutifully passed messages that seemed to look correct on to the keyserver. Had he coordinated with us, we could have given him the information on the servers used for testing. Williams' testing against the production servers killed the keyserver on May 21. Ninety minutes later, Verser succeeded in bringing the keyserver back online.

The following day, Verser posted a "Request for Patience and Cooperation" to the DESCHALL mailing list. He wrote that the outage had occurred because a user had been testing some buggy software against our production keyserver. He added:

> I have never claimed the keyserver was bulletproof. Maybe it should be more robust, considering it's only handling about a quarter-million requests per day, and only doling out around 300 trillion keys per day.
>
> You wouldn't believe the onslaught of strange and bogus data that comes to the keyserver. The keyserver manages to deal with most of it successfully. But neither the keyserver nor my ISP are bulletproof.
>
> If you support the DESCHALL effort, *please* do not test your code against the production (3.5 billion key/second) keyserver!
>
> Prior to the outage, I urged the person who took down DESCHALL to ask for advice on the mailing list. Here is an excerpt of his response. This e-mail was received about an hour and a half before the outage, while I was sleeping.
>
>> Even if you don't have experience with firewalls, I do, and it is a specialty of mine. The easiest way for me would have been to integrate the tunneling code directly into your code. Since you won't let me do that, I will have to write my own tunnel.

Verser added that the number of keys lost because of the outage was the equivalent of one Pentium system testing keys non-stop for a whole year. Verser stressed that he had no reason to suspect that Williams intended the project any harm, but his refusal to ask for advice and to wait a few hours for answers from the project coordinators had cost the project a bunch of keys. Though reports from many clients that never made it to the keyserver would simply be resent until the message got through, there was a loss of key reports in the system failure; some keys that had been tested would simply need to be tested again. It was a disappointing setback—and completely avoidable.

About the same time that work was being started on the gateways, discussion on the mailing list was focused on getting key-cracking clients that would run on as many machines as possible. University of Washington student Mike Heroux asked about creating a client in the Java Programming Language, so that anyone, on any computing platform would be able to run it.

Word of mouth and persuasion had worked well for us in getting new users to run DESCHALL clients on their systems so far. Usually, recruiting new system administrators to the project—or at least allowing someone to run the clients on their machines—was a simple matter of explaining the nature and importance of the project.

Sometimes, enlisting support got more complex. When Corey Betka at the University of Illinois at Urbana-Champaign approached a system administrator who had charge of over 150 machines, the system administrator explained that she'd be happy to run the clients on all of the machines in her care, if he was willing to help on a "project" of hers.

The system administrator—who was never identified by name—was collecting Beanie Babies and needed Ally the Alligator. So, DESCHALL would gain an additional 150 or so clients in exchange for a Beanie Baby. We had no idea how many deals like this were made to allow us access to additional machines to run the key-search clients, but what we do know tells us that the users who were running our clients and recruiting more volunteers were a dedicated lot. The willingness of users to engage in horse-trades like this with people who had much-needed computing power was testimony to the dedication of DESCHALL participants, and a critical part of growing our computing power.

19

Network

People working on our project—and competing projects—understood that we were not just trying to find a needle in a haystack, but that we were attempting to complete a huge computation—maybe the world's largest to date. Our "supercomputer" was unique because instead of having many processors tied closely together, it was made up of thousands of computers that were collaborating via the Internet. Sun Microsystems had been declaring, "The Network Is the Computer" since the early 1980s: the slogan was literally true in our case.

It's important to remember what the Internet connecting all of these thousands of computers was like in 1997. Ease of use greatly improved with the first graphical Web browsers from 1993 and 1994 led to rapid acceptance of Internet technology by mainstream computer users. At the end of 1994, there were roughly 38 million Internet users; 50.6 million by January, 1997; and 101 million by January, 1998. All of those users were supported by an ever-growing number of computer systems connecting to the Internet: just over 3 million in 1994 grew to 12.9 million in mid-1996, which grew to 19.5 million in mid-1997.[18] Internet service providers (ISPs) struggled to keep up with the demand in the face of this growth. Consider that to get a high-speed telecommunications circuit to connect a few hundred users to the Internet could take anywhere from thirty to ninety days to come online after an order was placed. With 1997's growth, that could mean that just in the time that the service provider waited for a circuit, the number of users would have increased anywhere from just under ten percent to twenty-five percent. In some places, the circuits simply could not be put into place fast enough to keep up with the demand for connections.

While the network providers struggled to accomodate the ever-increasing traffic, DESCHALL participants, like many other Internet users, felt the effects. Throughout the course of the project, for example, we had repeated reports that the keyserver was down. In most cases, the outage wasn't a problem with the DESCHALL keyserver or even its connection to the Internet but simply due to ISPs having more traffic on the network than the infrastructure could support. Internet traffic jams were unfortunately common during hours of peak demand. These hiccups were certainly a frustration for all involved, but the situation was not entirely unexpected, given the state of the technology available. When those exchange points were badly overloaded, what should have felt like one, seamless Internet became a visible amalgam of different networks; users on one part of the Internet could communicate with each other, but they would experience delays of minutes or even hours when trying to get data to users with different providers.

Connectivity problems were not limited to the exchange points. At the time, many homes and small offices relied on dial-up modems to get online. Dedicated circuits that provided continuous and high-speed Internet access were thousands of dollars monthly—and therefore only economical when serving dozens or hundreds of users. Reliable Internet connections were essentially limited to a rate of 28.8 kBps per second. Even then, most ISPs would hang up on connections that had gone idle for more than a minute or two—unless they were "dedicated dial-up" lines that cost five times as much as standard dial-up service. If your ISP hung up on you while you were reading a Web page, when you clicked a hyperlink, you'd need to wait for your system to dial up again, hope that you didn't get a busy signal, and then wait for the connection to be reestablished. Staying online for most individual users and even small offices was an arduous task. This was the Internet that we used as the basis for our distributed DES key-cracking computer.

20

Download

Friday, April 11
Yale University, New Haven, Connecticut

Jensen Harris knew what the problem was right away. He was reading a request for help posted to the DESCHALL mailing list by Josh Weage, a mechanical engineering student at Michigan Technological University. Weage had just heard about the DESCHALL project and decided that he wanted to help, so he went to the North American Cryptography Archive to download a client for his personal Pentium machine. He then decided to run the DESCHALL client software on one of Michigan Tech's Sun machines running the Solaris operating environment on a SPARC processor. Like he did earlier in the day, he went to the North American Cryptography Archive, downloaded a client for the Solaris/SPARC platform and started it up.

The program wrote some text to the screen, but didn't seem to be doing any work. Weage posted what the client wrote to his screen along with his request for help:

> DES Violation Client v1.0
> (C) Copyright 1997 the DES Violation Group
> gethostbyname: Error 0
> Obtaining keyspace from keyserver.des.violation.net.
> Error connecting to server.
> Waiting 2 minutes....

Having made the same mistake himself, Harris immediately recognized that Weage was not running the DESCHALL client but the DES Violation Group client. It was an easy mistake to make: both DESCHALL and DES Violation Group distributed their client software from the same site and both projects' clients had file names starting with "DES." Users just hearing about DESCHALL were bound to get confused. Most would simply hear about cracking DES keys, go to the client download site they heard about, and download the first package in the list that seemed to be about DES key cracking. Such users could hardly be expected to know that there were multiple projects.

Harris posted a response back to the mailing list, pointing out exactly which file to download, and then arguing the need for a separate archive strictly for DESCHALL client software. He made the same mistake when he first joined the DESCHALL project, and no doubt many others did as well. All of our efforts to promote the DESCHALL project could be rendered useless if people heard about our project and then started running DES Violation Group clients.

Harris didn't need to argue his case strongly—DESCHALL participants and coordinators all knew that he was right.

Saturday, April 26, 4:09 P.M.
The Ohio State University, Columbus, Ohio

Justin Dolske finished his program for handling client distribution for DESCHALL and decided it was ready for testing. He built the software around the basic requirement that it be easy for new clients to be uploaded by project coordinators and easy for eligible users to download the right clients. Part of the challenge for Dolske to teach the software to differentiate the eligible users from the ineligible ones. To address these concerns, we employed a combination of policy and technology.

Michael Paul Johnson's system for the North American Cryptography Archive apparently worked for its intended purpose, so Dolske built a system that would work in a similar, but not identical fashion.

Once Dolske's system was in place, users wishing to download the software first had to answer three questions, based on the questions for the North American Cryptography Archive but slightly modified by Rocke Verser:

1. Are you a citizen or national of the United States, a person who has been lawfully admitted for permanent residence in

the United States under the Immigration and Naturalization Act, or a Canadian citizen?
2. Do you agree not to export the DESCHALL client software in violation of the export control laws of the United States of America? Or, if you are a Canadian citizen, are you obtaining the DESCHALL client software for end-use in Canada by Canadian citizens, or return to the United States, in a manner permitted by Canadian law?
3. Do you assert that you have answered all of these questions truthfully?

If all three questions were answered affirmatively, the software then checked the visitor's IP address to determine if the computer making the request is actually located inside of the U.S. or Canada. Two different testing mechanisms were used: one based on the domain name system (DNS) that maps computer system names to IP addresses and the other based on the data from the registrars that manage domain names.

The test based on DNS was the simplest case. It had two parts that worked together. The first part was a DNS query by IP address to find the name of the user's computer. The second part was to query the DNS for the name from the first part to find its IP address. So, if I attempted to download the software from my computer called gatekeeper, Dolske's software would see that my computer's IP address was 206.98.200.180. The first DNS test would be to look the name up by address; that would determine that the name of my computer was gatekeeper.megasoft.com. The second part would look up the IP address of 206.98.200.180 to find it name, which would be gatekeeper.megasoft.com. This was a good test, because it would mean that two different sets of network managers (those who manage IP address space, and those who manage the domain name space) either agreed on the name and IP address of a system or independently delegated naming authority to the same person, which would also mean that person could name the machines whatever he wanted.

This DNS test would prevent "spoofing" attacks where managers of IP address space could make their systems look like they were really inside of the United States. For example, take a hypothetical company in France called *C'est Vrai* with the domain name of cest-vrai.fr, and a network manager (let's call her Renée) whose computer is called foo. Renée might go to the DESCHALL client archive running Dolske's software and answer "yes" to all three questions on the questionnaire. Dolske's software would perform its DNS lookups and see that Renée's machine asking to download the software is coming from the address of 192.168.254.11, which is known as foo.cest-vrai.fr. The second part of the test would show that foo.cest-vrai.fr really is 192.168.254.11. At that point, Dolske's software would then notice that the domain name is part of ".fr," the country code for France—and display this message for Renée:

> Error code: Export check 1 failed.
>
> Sorry. Based on your answers and/or other information, you are not eligible to download the DESCHALL client software. If you believe you have received this message in error, please mail us a note containing:
>
> - The time and day you received this denial (so that we may check why the server denied you access).
> - The hostname of the computer you were using.
> - A short note explaining that although you double-checked your answers on the form, you should still have access and why.

Being clever (and somewhat naughty, as network administrators frequently are), Renée might decide that Dolske's DESCHALL client archive software could tell she was coming from France based on the mapping of IP addresses to domain names, and that she could work around that since she had authority to change the address-to-name mappings. Renée would want change the address-to-name mapping so that 192.168.254.11 gives an answer that suggests it is in the U.S., perhaps foo.cest-vrai.us.

After making such a change, Renée could go back and try again. This time through, Dolske's software will look up the address and get foo.cest-vrai.us. The second part of that DNS test would try to find the address for foo.cest-vrai.us, which would not exist. Renée would be denied access a second time, seeing the same message with a different error code: "Export check 2 failed." The only way that Renée would be able to get around this system would be to find someone in the U.S. or Canada who was willing to modify their own DNS records to allow her machine to look like it was part of their network, and this was unlikely.

If double-checking in the DNS showed us that the computer was part of a domain in the U.S. or Canada, Dolske's software would present a list of clients that could be downloaded—but the user only had fifty minutes before the page wouldn't work anymore and the questionnaire and tests would have to be performed again. The domains that were obviously in the U.S. or Canada were ".edu," ".gov," ".mil," ".us," and ".ca." Other domains, like ".org," ".com," and ".net" could be from anywhere, so for these cases, Dolske's software would go to a second test: the registrar for Internet domains.

Return to the example of me trying to get the software from gatekeeper.megasoft.com. Since a ".com" domain could be from anywhere in the world, Dolske's software would query the domain name registrar to determine the address of the domain manager. The registrar would show that "megasoft.com" belonged to a company in Freehold, New Jersey, clearly inside of the U.S., and therefore grant me access. Anyone from Canada or any of the fifty states would be granted access. Not knowing whether allowing a download to a U.S. territory (like Puerto Rico or the U.S. Virgin Islands) counted as "export" under the regulation, Dolske decided to play it safe and deny requests from those areas. The rest of the world, of course, would be denied.

The system wasn't perfect, but since we were working on a project that would last months (not years) and people requesting the client software didn't know how we were determining exportability, it did serve its purpose. Plenty of people from ineligible areas of the world did try to download our client software and were denied. To this day, I know of no case where Dolske's software allowed a download incorrectly.

Sometimes, the system refused to grant access for users who were eligible to download, which is the reason why the form had a link on it to send email to Dolske and me. Anytime that we got a request from someone who said he should have access, we would investigate, using basically the same methods as the software—looking at the IP addresses and system names in the headers of the email request that we received for additional confirmation. If the user was incorrectly denied access, we would reply with a special password that could be fed to the system that would allow the user to download the software. The user could then answer the questions and put in the special password to be granted access to the client archive.

DESCHALL key-cracking clients were available for about thirty-six different platforms at this time, covering systems running Windows, OS/2, many different variants of Unix, as well as a wide variety of processor types. Within the week, new clients would be released for

Macintosh systems that were optimized for the PowerPC 601 and 604 processor models. Users could download any number of clients once they had been approved, so users with many different system types could go through verification once and then download all of the appropriate clients without needing reverification.

Thursday, May 1
Megasoft Online, Freehold, New Jersey

Justin Dolske and I both ran official DESCHALL client archives—using exactly the same software, and coordinated with each other so all of the DESCHALL client software for download was identical. For the first few days of operation, we advertised Dolske's distribution site at Ohio State as the "primary" and mine at Megasoft as "secondary," available in case the primary wouldn't work for some reason. After feeling comfortable that the system was working properly and seeing the kind of load that the system was putting on the underpowered computer Dolske was using at Ohio State, we decided to reverse roles, making my newer system at Megasoft act as the primary site, with Dolske's site acting as secondary.

We announced that the testing period was over, and the DESCHALL project had a new method for client distribution. While always grateful to Michael Paul Johnson for hosting our software on his North American Cryptography Archive from the earliest days of the effort, DESCHALL would now only be available through its own client distribution sites. No more confusion regarding which downloads were for which projects.

Although it was frustrating to have to glean participants from a pool of two countries, we felt we were making good progress nonetheless. Just as we were passing the point of having two percent of the total keyspace tested, however, one of our participants gave many of us cause to wonder just how much the cryptography regulations limited our efforts.

Sunday, April 27, 1:15 p.m.
University of Alberta, Edmonton, Alberta

Fourth year honors computer science student Howard Cheng was checking on our overseas competition and reading the SolNET project's Web site. Cheng read a figure that prompted him to send an urgent message to the DESCHALL mailing list: SolNET was claiming a spike in its search rate up to 1.5 billion keys per second, while our most recent statistics were showing only 1.16 billion keys per second. He correctly observed that SolNET had an advantage in being able to recruit participants from all over the world. SolNET seemed to have support from easily a dozen different countries, while DESCHALL, working under restrictions on cryptography export, was limited to the U.S. and Canada.

Not everyone was alarmed by the news. Justin Dolske observed that our rate had actually increased a bit to 1.25 billion keys per second. Guy Albertelli also noted that SolNET was reporting peak rates, while we were reporting the average number of keys tested over a 24-hour period.

Furthermore, the reason behind the sudden increase in SolNET key testing speed was no secret. It wasn't even the result of an effort to recruit a large number of new participants. A single site, a university in Taiwan, had started running 2000 SolNET clients over the weekend.

I quickly sent a message to the mailing list, reminding everyone that our project had only really taken off in April. Our firewall-traversing system had just been brought online and our client distribution system had just been improved. Our best answer to any challenges would be to recruit more people to run clients while the core DESCHALL team continued to improve the system architecture and develop faster clients.

But where would we find more clients? Looking at really large companies—AT&T, for example—reminded us that there were many tens of thousands of desktop systems available in the U.S. and Canada. If we could harness any significant part of that power, we could sustain our amazing growth for quite some time. Considering that the previous week had seen a 74 percent increase in speed up to a rate of 1.2 billion keys tested per second with 3000 participating machines, the potential seemed staggering.

By Tuesday, April 29, real comparisons were made between SolNET and DESCHALL, and an online dialog began as users from both camps debated the importance of the two organizations' statistics. One SolNET participant observed on its mailing list that DESCHALL was running at twice the speed of SolNET, that connectivity to our keyserver was better, and their effort had not grown significantly since their big boost over the previous weekend from the Taiwanese university. In fact, they had network and server problems that flattened their search rate.

Jeremy Bradley, a SolNET participant from Bristol University in the UK, asserted on the SolNET mailing list that DESCHALL had three "major weaknesses:" a single keyserver with an Intel 486 processor (to which he added, "I say no more," as if the computational power of the keyserver had anything to do with the rate at which the keyspace would be searched), difficulty with statistics (as evidenced, he asserted, by estimates being reported instead of precise figures—something we actually had not done since the middle of March), and our modest 3000 participating client machines versus SolNET's 15,000 clients. Bradley concluded with a prediction that SolNET would overtake DESCHALL within a week or two.

Adam Haberlach at Oregon State University responded with a message posted to the DESCHALL mailing list and copied to Bradley. Therein, Haberlach pointed out that our 486 keyserver was working just fine and that we were more interested in computing the right key than computing statistics. (In truth, our computers were busy searching keys, leaving us to talk about statistics; we could argue that other projects were busy computing statistics, leaving them free to talk about searching keys.) Haberlach also argued that we actually had closer to 5000 clients—but even with one third the number of clients, we were staying ahead of SolNET because our software was so much more efficient.

Whether DESCHALL was being hindered by being restricted to the U.S. and Canada had already been considered several times by this point, and we knew that many contest participants overseas would run the DESCHALL clients if they could. With competition from SolNET becoming more intense, more participants were more urgently needed. Because SolNET was able to recruit from the entire world—including North America, since EAR, like ITAR before it, applied only to *export* of clients; people in the U.S. could *import* any cryptography they wanted—we knew that we needed to continue in our recruitment efforts and improve our clients' speed even further.

Iowa State University student Mikael Brown wrote to the DESCHALL mailing list, musing that there would be some risk in a foreign project successfully answering the challenge. While the success of a foreign group like SolNET might show the futility of attempting to control crypto export to a technocrat, Brown noted that lawmakers had a tendency to view things from a very different perspective. He didn't need to elaborate, because many of members of the cryptographic community had long viewed legislatures and their restrictions on cryptographic products as irrelevant to the world around them. Many believed that the restrictions did nothing to stop the spread of cryptography to hostile entities abroad, but only prevented law-abiding citizens from being able to address global market needs. The danger Brown worried about was that if a foreign project like SolNET actually won the challenge, lawmakers might even attempt to *increase* the restrictions on cryptography.

What the future of this type of legislation would be was debated vigorously during the RSA contest. The discussion was not new to the contest, however, and was part of the arguments advanced in the Crypto Wars, the public policy debate on cryptography.

We could clearly see that U.S. policy disallowing the general export of cryptographic software was negatively impacting DESCHALL's ability to grow and helping SolNET to close the gap in performance.

Once again, we knew we urgently needed to get the clients running on more machines, and we were going to have to grow our ranks by appealing to a larger set of of users based in the U.S. and Canada.

Monday, April 28, 2:01 P.M.
The Ohio State University, Columbus, Ohio

DESCHALL coordinators were keenly aware of the challenges before us and were anxious to help our enthusiastic group of participants get more involved in the project. To that end, Justin Dolske composed a message to the DESCHALL mailing list simply entitled, "Things you can do."

Among the things Dolske enumerated were an improved front-end for the key-cracking clients, one that would allow for offline processing and better support for automatic stopping and starting so the client software would only work when no one was using the machine. This first request called for programmers who were familiar with the C programming language and who would be able to develp clients on Windows and Unix systems. The programmers did not need a particularly strong knowledge of DES since those requested components would not involve the key-testing parts of the client software.

Dolske also described a number of ways that someone might be able to enhance our progress graphs and statistical reports. Dolske was particularly interested in seeing someone develop a program that would allow participants to build queries of progress on particular groups of participants, such as all of the clients from a specific company or university, a feature that required more flexible system than the present Graph-O-Matic. Dolske finally called for people who could help us with publicity. We needed participants who could distribute press releases and explain the importance of the DESCHALL project to the media.

At this point in the project, DESCHALL's advantage of fast clients tied together with a capable and well-managed architecture was showing itself. We intentionally focused on recruiting new participants and enhancing core functionality instead taking the advice of some participants who wanted us to add new features and start cooperating with other projects. Rocke Verser decided—and other developers agreed—that having another set of client software, built and maintained by a different group of developers (from outside of the U.S.) for working with the DESCHALL keyserver would be a horrible distraction from what we did well, and that would probably not be possible to manage effectively. Dividing up the keyspace ahead of time and having, for example, SolNET search part, while DESCHALL searched another part, could be done if we had worked together from the beginning, but could have proved difficult once the projects were already up and running. Karl Runge's April 24 calculations about expected search time showed that the amount of return that we would have from working with another project was minimal, especially in light of the additional effort that would be required to pull it off.

21

Short Circuit

Thursday, April 10, 12:41 A.M.
Megasoft Online, Columbus, Ohio

Mike Heroux at the University of Washington asked a good question. He wanted to know why we didn't write a client in Java so that it could run on any type of computer, instead of spending so much time and energy in platform-specific versions. Answering on the DESCHALL mailing list, I discussed the tremendous performance advantage that some implementation strategies had over others. In particular, I pointed to the speed of the 32-bit Pentium clients versus to the clients for more powerful 64-bit processors. I observed that one of the best ways to increase speed was to optimize the client heavily for the specific processor in use, as Verser had done with the Intel clients. Another option I mentioned was the use of the "bitslicing" method described in a recent paper by Israeli cryptographer Eli Biham.

In typical key-search algorithms, the computer's processor will operate as it normally does—taking blocks of data and performing a series of functions on them. Various techniques were available for reducing the amount of work needed to test a key to see whether it was the right key to unlock the block of data. Biham's bitslicing key-search method took a rather different view of the problem.

Rather than treating a processor as a single component that works on numbers of a particular size (say, 64 bits), the processor is treated as 64 independent 1-bit processors. Thus, rather than the 64-bit processor changing from instruction to instruction through the course of the processing, each "independent 1-bit processor" performs the same instruction at each step, but with a different bit each time. This method can be used to reduce the total number of steps needed to test each key, resulting in a dramatic increase in speed. The end result is that roughly 300 instructions are needed to compute DES, where previously over 600 were used in other fast implementations.

This was the theory, at least. It would be up to implementors to prove just how much faster the technique would be.

Monday, April 18, 11:52 p.m.
Carnegie Mellon University, Pittsburgh, Pennsylvania

Darrell Kindred, a Ph.D. candidate at Carnegie Mellon University, had spent the weekend implementing Eli Biham's bitslicing method, because he was sure that if our clients could use the method, we could increase our key search. DESCHALL's code for the Intel processors was already the fastest of all known DES key search software thanks to Rocke Verser's tremendous optimizations for that platform. If Kindred could make bitslicing work for DESCHALL's clients on the more powerful 64-bit processors, we could get another boost in overall search speed. We would need this help if we were to stay ahead of SolNET, which was hard upon our heels. After performing some preliminary tests, Kindred sent his results to some DESCHALL coordinators. The results were very impressive.

Kindred's tests were run on three different machines. The first test was on a Digital Equipment Corporation (DEC) AlphaStation 255/300, with the 21064A microprocessor running at 300 MHz. As shown in Table 7, Kindred's bitslice client ran 106 percent faster than the available DESCHALL client for that system, 145 percent faster than the DES Violation Group client, and 177 percent faster than the SolNET client. Kindred also compared the speed of his software with bitslicing software from Australian programmer Matthew Kwan, who also wrote some fast DES functions.

Kindred bitslice software	1182k keys/sec
DESCHALL client	574k keys/sec
DES Violation Group client	483k keys/sec
SolNET client	427k keys/sec
Matthew Kwan's bitslice software	361k keys/sec

Table 7. Bitslice Performance on the 300 MHz DEC Alpha Processors

The next test was performed on a DEC AlphaStation 600 5/333 with an Alpha 21164 processor running at 333 MHz. Table 8 shows that increases there were also significant: 135 percent faster than the existing DESCHALL client, 175 percent faster than the DES Violation Group client, and 216 percent faster than the SolNET client.

Kindred bitslice software	2140k keys/sec
DESCHALL client	907k keys/sec
DES Violation Group client	775k keys/sec
SolNET client	677k keys/sec
Matthew Kwan's bitslice software	1720k keys/sec

Table 8. Bitslice Performance on the 333 MHz DEC Alpha Processor

Finally, Kindred tried his code on a different vendor's system, one from SGI, long known for their hot graphics computers. On an SGI Onyx with a 194 MHz R10000 processor, Kindred's bitslicing code ran 74 percent faster than the existing DESCHALL client, 157 percent faster than the DES Violation Group client, and 142 percent faster than SolNET's client. See Table 9.

Kindred bitslice software	1430k keys/sec
DESCHALL client	823k keys/sec
DES Violation Group client	555k keys/sec
SolNET client	589k keys/sec
Matthew Kwan's bitslice software	753k keys/sec

Table 9. Bitslice Performance on the 194 MHz MIPS R10000 Processor

DESCHALL already had the fastest clients, but Darrell Kindred found a way to get them to run even faster. These improvements were significant, because they showed that we could literally double our speed on these sophisticated 64-bit processors like the Alpha and R10000. Every 64-bit client being upgraded to a client using Kindred's software would be like recruiting another new machine. Pentium systems were still by far the largest contributor of processing power, but the improvements in the 64-bit system clients would have an immediate impact.

Meanwhile, DESCHALL started getting more publicity. *Communications Week* ran a blurb about DESCHALL in its "Security Monitor" column in the April 28, 1997, issue. Columnist Tim Wilson wrote:

> So far, about 2500 computers already are working on [cracking] the [DES-encrypted] message in a cooperative, "brute force" effort to try every possible key. More than 72 quadrillion possibilities exist, but the group is already trying 50 trillion keys a day, and more participants are joining in all the time, according to Rocke Verser, an independent consultant who developed the cooperative software in his spare time. At its current rate of growth, the group could decode the message in as little as two months, Verser said. Opponents of these export restrictions hope that if DES is cracked, the federal government will rethink its regulatory policies.

Wednesday, April 30
Virginia Polytechnic Institute, Blacksburg, Virginia

Blacksburg Virginia, which had been written up in "USA Today" in the summer of 1996 for being "the most wired town in the world," (as determined by the highest number of personal computers per capita) had a local paper, the *Roanoke Times.* That newspaper ran a full-length feature article on DESCHALL in its April 30, 1997 issue, covering the points we made in our press release and focusing on the contributions of local participants, including Alex Bischoff, whose picture was also printed with the article.

Also on April 30, DESCHALL got attention on the popular Macintouch Web site, a daily collection of Macintosh newsbits that thousands of people hit each day. In response to the recent release of new Macintosh clients, Jim Doolittle, a student at the University of Illinois at Urbana-Champaign tipped off the Macintouch site operators. The Macintouch site then included news about DESCHALL and provided a link to the client archive, starting a steady stream of new participants over the next few days, with the Macintosh clients being download several hundred times.

Thanks to the publicity—including media, advertising in online signature blocks, and word of mouth—not only were we searching more keys every day, but the mailing list also was becoming more active.

After the list bean receiving and distributing about thirty messages daily, some participants who really only wanted to see announcements started to ask for a way to keep up with the project without seeing everything that every project participant posted.

To offer some kind of reprieve, I opened a separate mailing list strictly for announcements relating to the project, including the release of new clients and important notes about any critical part of the DESCHALL architecture. The new list went live on April 30.

Beginning in late April, participants would occasionally write to the DESCHALL mailing list that they noticed delays when their clients would check in with the keyserver. Others were carefully watching the size of the blocks that the keyserver was handing to the clients for processing and reporting their observations to the mailing list. From these reports, some participants attempted to deduce the status of the keyserver. A few participants even started talking about "when DESCHALL adds a second keyserver," apparently assuming that DESCHALL would follow SolNET's footsteps in the addition of keyservers to its overall architecture.

Rocke Verser posted a message to the mailing list on April 30, addressing the question of the keyserver's status. "The keyserver is alive and well," wrote Verser. He continued:

> Up until today, it has generally been running at about twenty percent CPU. Today, it's probably between thirty and fifty percent. I won't know until the end of the day if that translates to more keys being checked or whether there were more packets being dropped on the Internet, requiring the server respond multiple times to the same request.)
>
> As I sit here, I can watch the "activity" light on the hub blink in step with the server's console log. There is no perceptible delay between the server receiving packets and the server responding to those packets.
>
> The server is easily handling over 4000 clients, and could comfortably handle 4000 more. Since I expect more than 8000 clients, plans are in the works for an additional server. But I emphasize, we don't need another server, yet!

> In an emergency, some minor changes to the keyserver could be made to increase the size of the average keyspace. Also, a "backup" keyserver is configured and can be brought online on very short notice if it becomes necessary.
>
> Some have asked why SolNet is already using multiple servers. I can't answer the question with any certainty. However I'll note that DESCHALL uses UDP [Unreliable Datagram Protocol] packets. SolNet by contrast uses TCP [Transport Control Protocol] packets. TCP has advantages for most protocols. But for this application, TCP is much more resource intensive and requires several low-level packet exchanges to accomplish what a single packet exchange in UDP can accomplish.

Questions about the keyserver subsided—at least for the time being.

Several hours before Verser's comment on the keyserver, our Australian participant, Andrew Glazebrook mentioned that he had a server available to use as a distribution point for DESCHALL software outside of the United States and asked if any of the official U.S.-based Web sites would provide links to his site for users outside of the U.S. and Canada. I quickly responded that while we could do nothing to stop him from doing anything that he wanted, we would not likely be able to provide links from any "official" Web sites to his. Even though we recognized the ultimate futility of trying to keep the software in the country, we had no intention of skirting the regulation while it was still in effect. Linking to a site where the software was distributed without regard to the regulation wouldn't be exporting the software, but a zealous prosecutor might find a way to argue that we violated "the spirit of the law," and none of us could begin to guess where that would wind up.

Glazebrook never did say how exactly he got the software, but we simply assumed that someone who could download the client, did, and then exported the software.

On May 1, Glazebrook posted to the DESCHALL mailing list that his site was distributing the DESCHALL client software for Intel processors. In that note, he explained his rationale a bit further: he was running OS/2 and SolNET simply didn't have an OS/2 client, so he figured that he could help other potential participants in the same predicament by having his own distribution site.

While some thought the idea was a good one, Darrell Kindred posted a note of caution:

> It seems quite possible to me that it will cause trouble for Rocke personally and/or the DESCHALL effort as a whole. Many of us hope to influence U.S. export policy through this contest, and I don't think it's going to help our case if the contest participants get portrayed as smugglers.
>
> If you're outside the U.S. and Canada, join the SolNET effort. We're all working toward the same goal.

As it turned out, the SolNet OS/2 client was released the day after Glazebrook posted the announcement of his Australian download site on the DESCHALL mailing list, but it was much slower than the OS/2 client we had developed. Ronald Van Iwaarden at Hope College in Holland, Michigan tried the SolNet OS/2 client once it was released, and reported that it tested 270,000 keys per second, while the DESCHALL client's 480,000 keys per second.

22

DESCHALL Community

Though DESCHALL had success in getting very fast client software developed, we still hand plenty of challenges to overcome. A real sense of community started to develop among DESCHALL participants as we worked through these obstacles and many users helped each other rather than waiting for one of the project coordinators to comment.

One situation where participants were able to help one another invovled tackling the issues faced by users who were connecting to the Internet via a dial-up modem. Reports filtered in from participants on dial-up machines. These participants found that their ability to contribute to the effort was hindered, sometimes significantly. If computers running the DESCHALL clients had to sit idle for hours, waiting for an active connection to the keyserver so it could report their activity and to get more work from the keyserver, any benefit realized by having extremely fast clients on those machines would be lost.

To combat this problem, many of the people running clients on these machines decided that they would simply remain connected for long periods of time. Unfortunately, this wasn't always an option. Many ISPs at this time oversold their capacity, since not all users who subscribed to the service would be online at once. As a result, an ISP could have 30,000 subscribers, with the capacity to serve only 15,000 at any one moment.

To prevent users from tying up service capacity unnecessarily by staying online continuously, ISPs would detect and disconnect users when the line they were using remaind idle for some period of time—anywhere from ten minutes to as little as one minute. Generally the rule of thumb was that the more oversold the ISP, the less time it would allow its users to remain idle online.

Colin L. Hildinger, Games Editor at *OS/2 e-Zine*, posted his solution to the problem. Hildinger configured his e-mail program to check for new mail every five minutes, thus staying under his ISP's timeout period while being permanently connected. By doing this, Hildinger's DESCHALL client was free to exchange information with the keyserver whenever necessary.

Milton Forte II, another OS/2 user, suggested the INJOY dialer, a program which would allow the system to use a "dial on demand" feature. Clients using this feature would be able to run DESCHALL normally even after their ISP dropped the connection because INJOY would automatically reconnect with the ISP when the client needed to communicate with the keyserver, just like the freeDUM program for Windows mentioned on the mailing list three weeks earlier.

Discussion on the DESCHALL mailing list on May 1 showed that freeDUM worked for only a few participants, leaving most to find more creative ways to keep their Windows systems in communication with the keyserver. Jason Gmoser, a participant from Florence, Kentucky, decided that he would just stay connected by having his system initiate a little bit of network traffic every five minutes. It worked well, but after having been connected for about seventeen hours, his ISP disconnected him. When Gmoser's system reconnected, he got a message that despite being been online all that time, he had only downloaded about five megabytes of data, thus indicating that he was holding the line, rather than actually using it the whole time. After getting that message from his ISP, he decided that he needed to find another approach.

Among DESCHALL users, there was no shortage of solutions to any problem—administrative or otherwise. Another participant working from home, Matt Clauson, quipped that Gmoser's problem could be solved by making a mirror of the entire software archive at Washington University—the largest collection of free software on the planet. Over a modem, the process would literally take months. In any case, it would prevent his ISP from complaining that he wasn't downloading enough during the time he was connected.

Still another participant working from home, Andrew James Alan Welty, suggested that the strategy of checking e-mail would be more effective if Gmoser mailed himself a twelve megabyte file. At 28.8 kbps, the upload to the mail server would take about an hour, and the download on the way back from the mail server should take another hour.

Still others suggested a simple change to a different ISP with sufficient capacity to handle a dial-up customer who was always online.

Participants frequently used the mailing list as a forum for all sorts of DESCHALL-related issues, including announcements of their own documentation and clever solutions to various problems before them. Stuart Stock, a systems and security administrator at Gundaker Realtors in St. Louis posted his "DESCHALL Linux Bootdisk Mini-HOWTO" on May 1.

Stock's Mini-HOWTO was a brief, technical document describing how to create DESCHALL "bootdisks" along the lines that Rocke Verser described in his April 10 message to the mailing list. The Mini-HOWTO document included configuration strategy hints as well as technical details, thus allowing system administrators to use, say, a network of fifty Windows 3.1 machines that had been booted from the DESCHALL Boot Disk to test keys all night long. The Mini-HOWTO even showed how the system would be able to run until it was time for work again, at which point the system would again reboot, but since the bootable floppy disk had been removed (immediately after starting the client), the system would start as usual.

The whole process was designed to be so discreet that system users never had any idea that the systems they were using during the day were hard at work on DESCHALL overnight. While this was no different from what system administrators were doing on university campuses and large companies around the country, it was a significant step forward because Windows 3.1 machines had not previously been able to run the DESCHALL clients. Stock's Mini-HOWTO showed participants how to bring DESCHALL to a whole new group of computers.

By this time, we had clients for nearly every platform in common use, including Windows 95 and NT, Macintosh, OS/2, Linux, and nearly every type of Unix workstation and server. Users of Sun's new UltraSPARC workstation and servers had just gotten a new client in the past week, bringing their performance from 640,000 keys per second up to 700,000 keys per second.

Users of the 64-bit systems, such as Sun's UltraSPARC, had not yet seen Darrell Kindred's work on the bitslice clients. Even as Kindred sent the incredible results of his testing to DESCHALL coordinators, users clamored for more improvement in the performance of the client software built for systems with non-Intel processors. While our clients were faster than those of the other projects, participants with expensive engineering workstations were disappointed to see their key testing rates compare so poorly to users with Intel-based PCs. The comparison wasn't really fair, since the Intel clients were so heavily optimized (see Chapter 13), but without any other means to compare performance, the users grew frustrated.

As Darrell Kindred continued his work on the bitslicing clients, he promised mailing list readers that he would explain in detail how the method worked, but not until after he finished his work and the new bitslicing clients were available for download.

23

Proposal

Wednesday, April 30
Capitol Hill, Washington, DC

Participants in projects like DESCHALL, SolNET, and DES Violation Group could easily get lost in technical minutiae regarding searching algorithms, architectures, and key sizes. While the DESCHALL team and our friends working on competing projects raced to see who could build the fastest and largest DES key searching system in the world, legislators in Washington were battling for the future of public policy on cryptography in the United States and abroad.

In the U.S. House of Representatives, the Security And Freedom through Encryption (SAFE) Act, designed to liberate cryptography, was making progress. It had been debated extensively and, despite the Clinton administration's objections that its electronic surveillance efforts would be hindered, the SAFE Act passed its first hurdle in a Judiciary subcommittee—unanimously. (See Chapter 7 for discussion of the bills and their rationale.)

The U.S. Senate version of the legislation was known as Pro-CODE. It was scheduled for a vote in the Commerce Committee on May 1. After seeing the SAFE Act's success, Conrad Burns, a Republican senator from Montana, led the bill's sponsors in delaying the vote for a month while they assessed their newly strengthened negotiating position.

"I think the administration sees the handwriting on the wall," said Burns' press secretary, Matt Raymond, to CNET News. Pro-CODE sponsors hoped that any changes that might further increase the bill's strength could be made by early June, at which point the bill would be ready to go to the full Senate for debate, and possibly even a vote. Helpfully, the Senate's majority leader, Republican Trent Lott of Mississippi, was a cosponsor of Pro-CODE.

Not everyone could read handwriting on walls, however. Less than two weeks later, an outline of a new bill by Senators John McCain (R-AZ), Bob Kerrey (D-NE), John Kerry (D-MA), and Earnest Hollings (D-SC) appeared on the popular Fight-Censorship mailing list. Known as the "Secure Public Networks Act," the bill was intended to give government officials access to information—even if encrypted. The bill required that all Americans using cryptography submit a copy of their encryption keys to a government-approved third party. If a government agency present the third party with a warrant for a given key, the third party would turn over that key to the government. Thus the McCain-Kerrey bill allowed the government some access to Americans' private information, while the other bills essentially restricted it.

DESCHALL was not only racing other groups participating in RSA's DES Challenge contest or fighting for a stronger replacement for the U.S. government standard for data encryption. For many participants and observers, DESCHALL was about asserting the right of the people to use cryptography freely. With the debate in Washington heating up, a new sense of urgency would drive our activity through the month of May.

24

In the Lead

Friday, May 2, 7:18 P.M.
MIT Campus, Boston, Massachusetts

Undergraduate student Ethan O'Connor started taking a look at SolNET statistics and noticed that their site was reporting 2 billion keys per second, with roughly 4000 hosts reporting in per half-hour. Using that as the basis for his calculations, he showed that SolNET was increasing in its overall search speed faster than DESCHALL. If progress for both projects remained constant, SolNET would pull ahead of DESCHALL in overall search speed within the week.

Meanwhile at Ohio State, Justin Dolske performed his daily checkup on our progress, as well as that of our friends at SolNET and DES Violation Group. We had not paid much attention to the DES Violation Group since the mid-April comparisons of our respective statistics pages that showed them lagging so far behind.

Instead of finding the group's latest progress statistics, Dolske saw a note that read:

> Due to lack of support, resources, and time to invest, the DES Violation Group has become inactive. With alternatives such as SolNET going more than ten times faster, our effort is just "wasting CPU cycles," to quote an e-mail we have received. We would like to thank all those who devoted time, effort, and CPU cycles to our cause.

Some of the DESCHALL participants speculated that with DES Violation Group falling so far behind and then seeing that SolNET was gaining on DESCHALL, DES Violation Group coordinators opted to throw their support behind the SolNET effort. We did think it strange that a U.S.-based project would not throw its support behind the other U.S.-based project—and front-runner!—but we were too busy testing keys to waste much time wondering why we didn't get the endorsement. Although we would have liked the DES Violation Group coordinators to recommend that their members join us, we knew that those members' CPU cycles were not lost to us: we were just going to have to recruit them more actively.

While we contemplated the potential impact of reducing competition in the DES Challenge, Nelson Minar at MIT drew our attention to an unwelcome development in the competition to recruit project participants.

Instead of waiting for the DES Challenge to be solved, Mike Driscoll, an engineering student at Colorado School of Mines, was organizing an effort to respond to a different contest—RSA's 1997 Secret Key Challenge for the 56-bit version of the RC5 cipher. (See page 44 for the announcement of the contests.) RC5 was not an interesting target because of what it was used for (most commonly, for securing Web transactions), but because attacks against RC5 were a useful way to show the relative strength of different key sizes. As a variable-key-length cipher, RC5 lets the implementor choose the desired key size between 40 and 128 bits. But there wasn't any particular importance to RC5; if 56-bit RC5 wasn't strong enough for a particular application, the implementation could easily be made 64 times stronger with a relatively simple configuration change to use 128-bit keys. With DES, no such reconfiguration was possible; it was hardwired to a 56-bit key—any variant to increase strength would require multiple keys, which introduced still other problems.

DESCHALL participants and coordinators were anxious to demonstrate the weakness of 56-bit keys in general, but we were disappointed that Driscoll was proceeding before waiting for the DES Challenge to finish. Although the disappearance of DES Violation Group would make more participants free to participate in either DESCHALL or SolNET, the appearance of Driscoll's group to break a 56-bit RC5 message by brute force, known as Bovine, would pull some potential participants away from the DES Challenge and to the RC5 Challenge.

Although DESCHALL was still going strong, Driscoll's project raised a couple of concerns. First, if we were to really prove the fallibility of DES, we needed to crack it quickly to show that DES was too weak for practical use. A secondary problem would be the way that an 56-bit RC5 crack taking place before a DES key crack would play in the media. Imagine the headline, "56-bit Cryptography Cracked; Actual Standard Still Safe." While we could probably argue that the Bovine project demonstrated the weakness of 56-bit keys, the point of our project was to demonstrate how the actual standard itself was vulnerable to brute-force attacks.

DESCHALL participants understood these issues, so we had no concerns about them switching to the other contest. We did worry that potential participants would hear about the Bovine project and choose to participate in it instead of joining the ranks of DESCHALL.

25

Recruiting

Friday, May 2
Megasoft Online, Columbus, Ohio

While it was tempting to worry, it was too early to tell whether the Bovine groupd would have a negative impact on DESCHALL, and frankly, I had other, more urgent matters to address. To explain the purpose and importance of the DESCHALL project and to help people get started participating, I began to compile a list of frequently asked questions and put them into a document along with the answers. Participants who joined the effort early tended to understand cryptography and computing technology pretty well, but as we got more attention through the media, we knew that we could not count on participants just joining us to have the same technical background. The "Frequently Asked Questions" document was intended to allow people who had heard about DESCHALL or the RSA DES Challenge to find out more information about the effort, and decide if they wanted to participate and how they could do so.

Finally, as the work week came to an end, I finished the first release of the document and put it into place.

During our discussions about the disappearance of DES Violation Group, the acceleration of SolNET's key testing rate, and the possible implications of a premature effort to recruit people to the 56-bit RC5 effort, something was quietly happening: DESCHALL experienced a significant increase in its overall key testing rate.

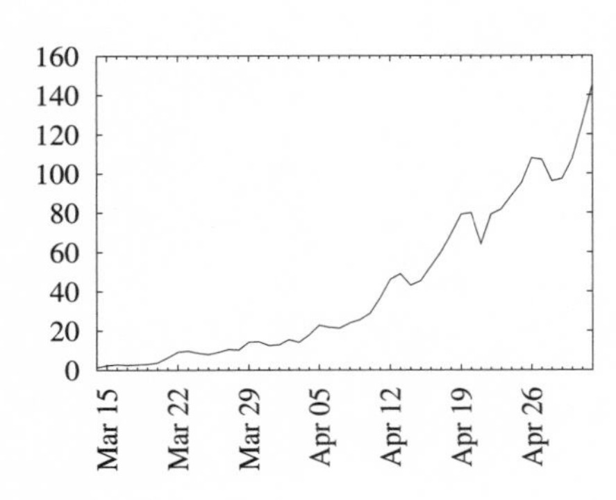

Fig. 6. Keys Tested per Day, in Trillions

The morning of May 3 turned out to hold good news for us. The release of the of the previous day's statistics showed that we had started testing the keyspace much more quickly.

Nelson Minar also started paying attention to the standings of various teams of participants, seeing how MIT was doing by comparison to some other universities.

Around the same time, Colin Hildinger compared our most recent statistics with SolNET's. While we had searched two-tenths of one percent of the DES keyspace the previous day, SolNet managed to search two-and-a-half-tenths of one percent in a nearby twenty-four hour period. Because of delays in getting statistics calculated and published, we wouldn't know until the following day—May 3—how exactly we would compare. But it was obvious that DESCHALL and SolNet were running neck and neck.

Keys	Clients	Domain / University
17.0 Trillion	393	uiuc.edu, University of Illinois at Urbana-Champaign
7.3 Trillion	372	orst.edu, Oregon State University
5.9 Trillion	191	mit.edu, Massachusetts Institute of Technology
5.7 Trillion	203	mtu.edu, Michigan Technological University
5.4 Trillion	187	cmu.edu, Carnegie Mellon University

Table 10. Comparison Among Universities, May 2

Although SolNET was closing on our key testing rate, we were still far ahead in terms of total keys tested and our clients were so much more efficient than the SolNET clients. There just wasn't any way for either project to back down at this point—SolNET users outside of the U.S. couldn't get to our software, and DESCHALL participants didn't want to run slower key testing software. The only option was to drive forward, turning up the heat on one another in the process.

Meanwhile, IBM's chess-playing computer, Deep Blue, was slated to play against the world's highest-rated chess player, Garry Kasparov, in another week. (Deep Blue would go on to win in a highly-contested match, the first time that a computer ever defeated a reigning chess champion.) Nate Boyd, from MIT's Computer Graphics Group, suggested that we try to recruit Deep Blue for cracking keys. Boyd was joking, but in 1998 the Electronic Frontier Foundation, an online civil liberty group would fund the development of a DES key-cracking machine called "Deep Crack," a special-purpose computer built for the sole purpose of cracking DES keys faster than software ever could.

Instead of trying to recruit Deep Blue, Justin Dolske tried to get some attention for DESCHALL by submitting an article to comp.os.linux.announce, the same newsgroup that had carried the RC5 announcement that had so recently caused a stir among DESCHALL participants. Given that the moderators had allowed Mike Driscoll's article about the RC5 Bovine effort to be released, Dolske thought that interest in cryptography would be high enough to warrant some publicity for DESCHALL. Dolske's announcement, just like Driscoll's RC5-56 announcement, was simple and to the point: what the DESCHALL project is, why it is important, and how to join.

The newsgroup moderators didn't seem to agree, and rejected Dolske's post, on the grounds that they just allowed something about the RC5 effort recently. Even more strangely, after rejecting Dolske's message, the same moderators allowed for *another* post for an RC5 effort. The moderators never answered questions about their decision, leaving some to conclude that there was some bias against DES (or in favor of RC5) efforts. This little mystery remains.

I could think of no reason to reject Dolske's announcement about DESCHALL. DES was a real standard, even used in the Linux operating system. RC5's use was not nearly as widespread, the 56-bit key size was not a typical configuration for RC5, and there was no direct connection between the Linux operating system and RC5. The decision to reject Dolske's post seemed completely arbitrary, and the piece might have appeared if a different moderator happened to see the submission.

After no small amount of concern about if (and when) SolNET was going to surpass our key search speed, we started taking a harder look at the numbers in order to get a better grasp on the progress being made by both projects.

SolNET's progress statistics were based on the number of clients that checked in with the keyserver during a given half-hour period. Some DESCHALL participants thought that all of SolNET's clients were checking in every half-hour and thus assumed that the statistics reflected the combined efforts of the entire project. On May 4, Rocke Verser pointed out that computers running the client software reporting their progress in that half-hour window were most likely reporting more than one half-hour's worth of work; many other clients were busy testing keys, but did not have any results to report back to the keyserver during that time. Applying the same metric to our own progress, Verser wrote, "In the last 60 seconds, 180 DESCHALL clients reported 122 billion keys tested. (Wow! We're testing over 2 billion keys per second with only 180 clients!)"

As it turned out, the keyserver numbers were really only useful for showing how much of a load the keyservers were maintaining. SolNET's keyservers were taking roughly two hits per second from their clients, whereas the DESCHALL keyserver was taking about three hits per second from the clients.

Looking at total project speed for comparison, Verser could see that SolNET was testing just under 2 billion keys per second, whereas DESCHALL was testing just over 2 billion keys per second.

Verser correctly concluded that SolNET had *more* clients—more than three times as many computers running their client software—but DESCHALL had *faster* clients.

Unfortunately for SolNET, their numbers turned out to be less impressive then they had intially appeared. Fredrik Lindgren confirmed on SolNET's DES-Announce mailing list that some machines in Finland were pretending to run SolNET clients, but actually were not. Basically, someone examined the protocol running between client and server in order to figure out what exactly to send to a keyserver to impersonate a client. Having figured out the protocol between client and server, the malicious client would simply send a "block finished, but key not found" message to a SolNET keyserver, without ever having tested a single key. The goal of such an attack was to make the key-server incorrectly record blocks as being tested, potentially preventing the real key from being found.

This was exactly the kind of thing that everyone involved in distributed computing efforts worried about. No one really gets anything out of an attack like this, but if it's easy enough to do, some idiot somewhere in the world is certain to try it. Lindgren and other SolNET coordinators had their hands full trying to assess the damage and building a plan for recovery, so they did not bother trying to figure out who was behind it.

Many DESCHALL participants began to ask what we were doing to prevent this kind of attack on our systems. I raised the issue privately with Rocke Verser and Justin Dolske; we talked about the problem and some of the defenses that we had in place, but never directly answered the questions presented by general participants. If the attackers were reading DESCHALL's lists, we didn't want to give them any clues about what our weak points might be.

SolNET would not be completely undone by this attack, but it was a setback. The bogus reports came from a particular network on the Internet, allowing for all of those reports to be discarded, ensuring that those keys would be tested by another client later. If any legitimate testing had been done, those would be lost, but the integrity of the project was critical and it was better to retest a key that had been tested once than to miss a key that was falsely reported as tested.

26

Threats

While SolNET was recovering from its attack, the DESCHALL team turned to other issues. Having enabled the basic mechanism for searching for DES keys on thousands of machines, of dozens of types, from all over the United States and Canada, our project had become quite visible, and it was possible that the same attackers who targeted SolNET might try to foil us. We did not want just to worry about what kinds of bad things could happen; we wanted to anticipate the most likely threats so that we could methodically address them.

Sunday, May 4, 2:20 A.M.
University of Texas at Austin

As DESCHALL participant Seth Johnson read CNET news, he found an article that discussed a math bug in the Intel Pentium Pro microprocessor. The article specifically described a problem that could cause some calculations to be done incorrectly. Users wouldn't generally notice the problem; it was the kind of obscure thing that only people performing heavily numerical scientific work would ever encounter in a meaningful way.

Ordinarily Johnson would not have been interested in the problem, since his Macintosh did not use an Intel processor, but now that he was participating in DESCHALL, he had cause for concern since so much of what DESCHALL was accomplishing was on Intel processors. If the bug affected the way that DESCHALL ran on Pentium Pro processors, we might need to throw out huge amounts of completed work to calculate those key blocks correctly. A problem like that would be a terrible setback—possibly setting us back more than a month.

Since the new problem affected relatively few applications, Johnson didn't want to create panic, but he did want to know whether this was a serious concern for DESCHALL.

Fortunately, the integrity of the DESCHALL results was not affected. As it turned out, the bug was limited to the floating-point unit of the microprocessor, where operations on "floating-point" numbers—numbers with decimals like 3.141592654—would be handled.

Since computing DES keys was done by working with 56-bit binary numbers and never performing any division operations that could leave a remainder, DESCHALL software had no need for floating-point operations, using only integer—"whole number"—operations.

While that particular processor bug did not affect DESCHALL, it did give us cause to think about some potential setbacks. Among those concerns were the possibility of long-term network outages making the keyserver unreachable and our potential vulnerability to attacks from malicious clients reporting bogus results back to the keyserver.

In general, a network outage wasn't likely to be a big problem. A DESCHALL client would work through the keys assigned by the keyserver. Once it was finished, it would attempt to report an update to the keyserver and to request a new block of keys to test. If the keyserver could not be reached, the client would simply resend its request at a later time. If some of the clients have to wait for a few minutes, this isn't a big problem, but if the keyserver is offline for a longer period of time, the impact grows quite a bit.

Assume for the purposes of illustration that the typical DESCHALL client needs to talk to the keyserver every other hour. Also assume that the other participating clients were started at evenly-distributed points throughout any two-hour window. A network outage of fifteen minutes would mean that the keyserver would be unavailable to respond to clients during a time when one-eighth of the total number of clients needed to check in. A thirty-minute outage would idle one-quarter of our clients for some period of time. After two hours, virtually the entire project would be at a standstill, with almost all clients waiting to connect to the keyserver. Given our key testing rate at the beginning of May, that would mean that 7.2 trillion keys per hour would not be tested.

In reality, clients would not be started to evenly, but the assumptions above demonstrate a critical point of the actual project: with less than a half-day of unavailability, the world's fastest distributed computer would grind to a halt.

The concern about malicious clients reporting bad results back to the server was one that came up on the Coderpunks mailing list, an offshoot of the Cypherpunks list. While Cypherpunks dealt with political, economic, and cultural issues as well as technical concerns about cryptography, Coderpunks focused specifically on the implementation of cryptosystems. Many Coderpunks readers were watching all of the RSA Secret Key Challenge contest projects, and more than a few were participating in one of the projects.

After a discussion of various ways to deal with the problem, Coderpunks subscribers concluded that the only sure way was to verify each of the test results. Trying to find the right key among 2^{56} was hard enough. Checking each key twice was less appealing. However, if the author of a malicious client was smart enough, there was no other sure way of knowing that we were dealing with good results.

DESCHALL clients did have some safety features built in, though, and thus was not wholly vulnerable. Although we did not reveal this in detail to the DESCHALL mailing list, DESCHALL client developers knew that DESCHALL clients weren't going to be too easy to spoof. "Half-matches," (see Chapter 13) for example, were reported, and some number of half-matches would be expected over a set of a certain size, so if Rocke Verser noticed a change in the expected half-matches, that could indicate that someone was reporting bogus results to the keyserver. Still, in spite of the precautions that we had taken, a determined attacker would be nearly impossible to stop. We really were just hoping that we could make it more difficult than it was worth.

Another problem that could arise would be if a client could somehow prevent a "key found" message from being reported back to the server. The client could, for example, get a key block, disconnect from the network, and test the keys offline. Normally, if the client found the key, it would send a message back to the keyserver. A malicious participant might prevent his client from giving the "key found" message to the keyserver, allowing the malicious participant to keep all of the prize money.

Of course, after the fact, the keyserver would show that the key had been given out for testing, but the results were never returned, a fact that would raise more than a few red flags. A little investigation would turn up where the key came from—though there is probably some legitimate question about whether that kind of evidence would be enough proof for RSA to refuse to award all of the money to the client operator.

While Rocke Verser, Justin Dolske, and I wrestled with these potential problems, some of our participants were dealing with issues of their own.

Tuesday, May 6, 2:22 A.M.
Western Michigan University, Kalamazoo, Michigan

A student named Jay, going by the alias "A Psychedelic Psychopath," was working with a few other students to run DESCHALL clients on some fairly powerful computers. Having access to two groups of Sun SPARC 5/20 systems they decided to put some of those machines' cycles to work for the project.

Then two system administrators noticed their efforts. One killed off the programs, claiming that the DESCHALL client programs were putting too much load on the systems, and the other disabled the students' accounts and reported Jay and his friends to the dean of students. Nothing serious happened to the students, but the incident did get us thinking about how to deal with system administrators who ran the powerful systems that so many participants wanted to use for testing keys.

Prompted by the report of what happened, Colin L. Hildinger asked if it might make more sense to give the program a more academic-sounding name. Some participants thought this was a good idea and favored giving them names like "lab3," at least in departments where students were given assignments to write programs. In that case, instead of system administrators seeing that students were running DESCHALL software on school systems, the clients would look like homework.

Vincent Fox, a system administrator at Georgia Tech, weighed with some helpful advice to anyone who might be trying to find a way to work around system administrators. "Better to go to your admin," wrote Fox to the DESCHALL mailing list, "and enlist their aid in this project in the first place."

"I know of very few who wouldn't actively help out if asked to be a part of the project rather than having it inflicted on them. You might find that they can in fact toss more resources and butt-covering your way than you can possibly imagine. We [administrators] usually have access to more hardware than you [students] do, and [have] less accountability."

Fox was right, as we would see repeatedly. People taking the time to enlist the help of administrators were much more successful in getting more systems to be used for key testing than those who tried to be more covert. Nevertheless, getting cooperation was simply not possible everywhere. Some organizations would actively encourage participation in this kind of project, while others wouldn't allow it for a moment. In general, large, highly formal organizations like a bank would not allow DESCHALL or similar projects, while more dynamic organizations with highly-empowered employees would encourage it.

Most were somewhere in between, and DESCHALL participants quickly learned just where.

Tuesday, May 6
IBM Almaden Research Center, San Jose, California

That cryptography had gone mainstream was clear, thanks to regular news stories on the progress of the DES Challenge contest and public policy debates. IBM was about to test just how closely the media wanted to follow developments in the field.

IBM's public relations machine put the word out on a paper that IBM computer scientists Miklós Ajtai and Cynthia Dwork presented at the Association for Computing Machinery's Symposium on Theory of Computing. Scientific papers are presented at computing theory conferences all the time without fanfare, but the implications of this particular paper were apparently too great for IBM to resist drawing greater attention to it. The result was a lesson for DESCHALL coordinators that getting the story wrong and missing the point were two very real possibilities if the information did not get presented to the press just so.

Prompted by the public relations behind the Ajtai-Dwork paper, *PC Week Online* ran a story that began, "Researchers at IBM's Almaden Research Center Lab in San Jose, Calif., claim they have discovered public key encryption that is uncrackable, solving a problem that has defied mathematicians for 150 years."

The Ajtai-Dwork system was not "uncrackable," as the media would claim it was, but it was an interesting new system for encrypting data based on a different mathematical basis from those used in other cryptosystems. The "uncrackable" part of the story came from a mathematical proof included in the paper that demonstrated that a random attempt to guess the key is the equivalent of the hardest possible case. In other words, with the Ajtai-Dwork system, it's not possible to pick a weak key by accident, thus making decrypting a message easier than it should be.

Using a mathematical proof of security was a bold step, very different from how cryptosystems are usually presented to the scientific community. The way that it usually works is that someone will come up with a new cryptographic algorithm and publish it in a paper, so others can study it for weaknesses. Over a period of several years, the algorithm will be attacked in various ways, and cryptographers will see how resistant the algorithm is to the attacks that are used against it. Over that period of time, confidence in the algorithm will increase, because the longer than an algorithm is studied by the world's smartest cryptographers without a break, the less likely that such a break will be found before computer technology progresses to the point where messages encrypted with the system could be broken by brute force.

This method tends to work pretty well, albeit slowly, and has resulted in the development of some very good algorithms, including DES. The problem is that if someone were to find a brand new kind of attack against a technique used in well-established algorithms, it could be a nasty surprise for everyone. New attacks that work against real ciphers don't come to light every day, but they are discovered frequently enough to keep cryptanalysts all over the world looking for more. Sometimes, systems are in use for years before a practical attack is found against them. On a few occasions, systems being presented at scientific conferences have been broken while the paper is still being presented. Building and breaking ciphers is hard work and full of uncertainty. We have many systems that are "probably secure" and many others that were "probably secure" until someone figured out how to break them.

Using a "provably secure" cryptosystem—one whose correctness was supported by a formal mathematical proof—was another matter entirely. In a provably secure system, each possible way to decrypting the ciphertext is represented mathematically. Then proving which is the easiest way to deduce the corresponding plaintext will prove the actual strength of the cipher. This is somewhat like proving the strength of each link in a chain. Finding the weakest link will prove the strength of the chain as a whole.

The problem with provably secure systems is that they have to make a lot of assumptions about the world around them. For example, someone who accidentally lets someone else discover a key would open up an avenue of attack that would be very practical in the real world, but would not ever make it into a mathematical model. Provably secure is a long way away from unbreakable.

By the time the media was putting the discovery in front of people, IBM had invented an unbreakable cryptosystem. Not provably secure, but unbreakable. DESCHALL coordinators would take seriously the lesson of how esoteric scientific discoveries can take on lives of their own once in the hands of the media. While wanting to use the media to get an important story to potential participants, we did not want the attention at the expense of accuracy in what we were doing. When it came time to talk to the press, we would take great care to ensure that we were being both as precise and clear as possible.

The lesson came at an apropos time: media coverage of DESCHALL was increasing.

MIT's student newspaper, the *Tech*, carried an article about DESCHALL and the efforts of MIT students there to help with the search for the key. The next day Nelson Minar told the DESCHALL mailing list that, within fifteen hours of the paper's release, 120 new machines at MIT joined in the DESCHALL project—an increase of fifty percent.

Three days later, Michael Nelson informed the list that the Utah State University student paper, the *Statesman*, ran a front-page article on DESCHALL. Although the article jumbled significant technical details, they correctly quoted Nelson: "I think we should push past BYU [Brigham Young University]. It is a good way to show that BYU isn't the only university in the state."

As more students started to join the efforts going on at their schools, rivalry began to grow—and this competition would work to our advantage.

When Nelson Minar reported the increase in activity due to the article in MIT's the *Tech*, he also predicted that MIT would overtake the University of Illinois at Urbana-Champaign (UIUC) in the total number of machines participating and perhaps even in number of keys tested per day.

Forty-five minutes later, UIUC Unix systems engineer Joe Gross responded with a post back to the DESCHALL mailing list. "We'll have to see about that," wrote Gross, before describing the fifty high-end UltraSPARC machines from Sun Microsystems that had just been brought online and were about to start running DESCHALL clients. Gross added, that "once finals end next Friday," UIUC would start running DESCHALL clients on 300 high-end workstations.

Back at MIT, Nate Boyd wrote, "That just ain't fair. The NCSA is definitely contributing big!" NCSA—National Center for Supercomputing Applications, located on the UIUC campus—had lots of horsepower available, and was, coincidentally, where Marc Andreessen and friends wrote Mosaic, the first graphical Web browser. As it turned out, Gross was not talking about getting any of the supercomputers to run DESCHALL, but the idea that the university was making such a big contribution was enough to turn a little heat onto the healthy rivalry that was already driving university students and staff to get more systems running DESCHALL clients.

Adam Haberlach at Oregon State caught the spirit of threatening the use of supercomputers and joked to the DESCHALL mailing list, "Don't make me try and get time on our Oceanography Department's CM5. Or the CS department's Maiko." (For all of the posturing that students from schools with powerful supercomputers were doing, the simple fact was that supercomputers really wouldn't help much, even if we did have DESCHALL clients that they could run, as described on page 58.)

MIT undergraduate Will Koffel observed that the MIT wasn't really doing anything to support DESCHALL or the users trying to find places to run the clients. He wrote, "I'd like to see UIUC test 16 trillion keys a day with dorm room computers!" Koffel noted that MIT had some die-hard DESCHALL fans on campus, who were sneaking into their labs and firing up clients on all the SPARC 20s and Pentium machines they could find. Koffel proclaimed, "We'll take the peak yet!"

Jeff Gilchrist was coordinating DESCHALL activity at University of New Brunswick (UNB) in Canada. He had spoken with the faculty of Computer Science and Computing Services about DESCHALL, and was given permission to run some of the clients in a few different department labs. He had even managed to get some of the professors to run it on their machines. While not "sponsoring" Gilchrist, the university administration were certainly aware of his work and approved of his running the clients.

Adam Haberlach was in a similar situation at Oregon State. Though not formally backed by the University, individual departments were giving him some support. Many students and staff members at Oregon State were participating, a few labs were running the clients, and some Web pages were describing the efforts on campus and explaining how others could join in. Among the departments participating, the Business Department was actually the biggest contributor. Oregon State's rankings went from first to second in early May, and then lower, when the DESCHALL supporter there took a trip to a big trade show in early May. Haberlach added with a smile, "We'll be back."

Like Haberlach and Gilchrist, Benjamin Peterson at Notre Dame wasn't getting direct support either, but had managed to coordinate things so that there would be no interference with people using the machines. He had 170 Sun UltraSPARC 1 machines and ten SGI machines working on the project. Between 8:00 A.M. and midnight, if someone logged into the console, the DESCHALL client would be killed. When no one was logged into the machine's console, or outside of those "daytime" hours, Peterson's program would be sure that no one was logged into the machine and working remotely. If the machine was being used that way, his program would stop the DESCHALL program—putting it in a kind of suspended animation. If the users on the machine all went idle for more than ten minutes, his program would tell DESCHALL to continue, picking up right where it left off. With some additional checks to suspend and to continue the process depending on the load of the machine, Peterson's contribution to our efforts was significant. His program was also so well designed that no one would have any reason to complain about the program using too much system resources.

Peterson wasn't the only participant who built additional software to manage the clients on a large number of machines. At the Bowman Gray School of Medicine, system administrator Dave Ahn wanted to run the DESCHALL clients on his SGI systems with their fast 64-bit R10000 processors. On May 9, he released WFU, the Watch Fork Utility for DESCHALL.

As summer grew near and students began working on year-end projects, daytime activity on the machines Peterson was using would tend to peak at about twenty of the Sun machines active at any point. After the year-end projects were finished, most points of the day had between fifty and eighty machines active.

Beyond that, a few more students and staff were running the clients in their labs. Unfortunately, support at Notre Dame didn't grow like it did in some other environments, and as the summer rapidly approached, many of the students that were participating from all over the country, including Benjamin Peterson, were leaving campus for home.

While some DESCHALL participants were preparing to head home for the summer, others continued to join the project. As the first full week of May wound down, the DESCHALL keyserver accepted a report of some keys tested from an unlikely contributor, whose report of some 30 billion keys tested earned a subtle entry toward the end of the DESCHALL status report for May 8.

Keys	Clients	Domain
30 Billion	1	sollentuna.se

Table 11. Surprise Entry in May 8 Report

Friday, May 9, 4:20 P.M.
Sollentuna, Sweden

SolNET DES project coordinator Fredrik Lindgren grinned as he read the DESCHALL mailing list and saw the question, "Hey, don't they have their own project?" The day before, Lindren went to Andrew Glazebrook's Web site in Australia and downloaded a copy of the DESCHALL client. Lindgren posted a friendly note to the DESCHALL mailing list explaining that he simply wanted to see how the DESCHALL client ran and wished us well.

Some DESCHALL participants suggested that it made sense for someone from SolNET to check our progress so that they wouldn't test the keys that we've already tested. Someone else then suggested that we should "respond" to the "offense" by blocking requests to the keyserver from their domain and changing the way that we assign blocks of keys to test.

Justin Dolske sardonically responded to the suggestion of blocking SolNET's communication with our keyserver with a mock-paranoid message posted to the DESCHALL list. He wrote:

> That wouldn't help. Sollentuna is run by the NSA [National Security Agency], and they're already using alien technology to read Rocke's brainwaves. In fact, part of the [non-disclosure agreement] the rest of the developers have signed requires us to wear a hat wrapped in tinfoil, to try and protect our thoughts.
>
> Additionally, we are starting to suspect that most of the domains "helping" DESCHALL are, in fact, fronts for the NSA, FBI, BATF [Bureau of Alcohol, Tobacco, and Firearms], PBS [Public Broadcasting System], and UN [United Nations]. In the next few days, we will be locking out all of the currently participating domains, in order to keep out these secret government agencies.

Many others responded more seriously, pointing out that we're really all working for the same goal. Rocke Verser summarized the issues nicely in his response. He wrote:

> SolNET has placed us twelve seconds closer to the solution.
>
> It's not my concern how they got the software. I suppose it's the concern of the FBI. All I know is they didn't get it from me. The U.S. border leaks like a sieve with respect to cryptography products. (Another example of how U.S. law hurts U.S. industry: the products get to foreign soil anyway but royalties don't come back to the U.S.)
>
> SolNET has announced that they preassigned their "master" keyblocks in a random order, and that they are distributing their keyspace in the order preassigned.
>
> DESCHALL does not issue its keys sequentially, but it's not entirely random, either. How DESCHALL assigns and processes keyspace is subject to the Non-Disclosure Agreement our developers have signed, so further comment is inappropriate.
>
> As far as blocking their requests—I have no intention of blocking anybody's request unless they somehow abuse the keyserver.
>
> As far as disallowing their access to the mailing list—I trust that Matt will continue to allow everybody to read our mailing list! (I read SolNET'S mailing list.)

SolNET and DESCHALL are working competitively, toward a common goal. I certainly hope that a DESCHALL client finds the key. But if the Probability Gods gave me the choice of letting SolNet find the key tomorrow or DESCHALL finding the key in forty-four weeks, I'd let SolNET find it tomorrow.

And finally, the SolNET organizers are a class act. They have never said or done anything distasteful or disparaging towards DESCHALL.

They were forthcoming when they discovered their server had been [attacked]. (They could have pretended it wasn't happening; they could have continued to claim a highly inflated keyspace rate, but they didn't.)

As I said in my posting to their mailing list a few days ago: "If U.S. export laws weren't what they are, I suspect we would be collaborating."

The campus rivalry that had been stirred up continued over the course of the next several days—this time with UIUC and Georgia Tech. Statistics showing progress on May 10 put UIUC at the top of the heap.

Keys	Clients	Domain
22.28 Trillion	543	uiuc.edu
22.21 Trillion	683	gatech.edu

Table 12. UIUC Ahead of Georgia Tech, May 10

Georgia Tech's Perry Minyard found another important entry in the logs. A group of four machines working in the network address space of 130.207.0.0 did not have names in the reverse DNS—the mapping of IP addresses to names. Hence, they showed up as IP addresses in the statistics, and not part of the record for gatech.edu. Looking at those statistics together produced a different result. Minyard proudly proclaimed Georgia Tech the site leader in keys tested.

Dave Terrell and Jay G. Lickfett, both at UIUC, then noticed another important line in the same DESCHALL progress report. A client

Keys	Clients	Domain
22.30 Trillion	687	gatech.edu and 130.207
22.28 Trillion	543	uiuc.edu

Table 13. Georgia Tech Ahead of UIUC, May 10

was participating from the network 128.174.0.0 without reverse DNS. The 128.174.0.0 network belonged to none other than UIUC. Terrell and Lickfett recalculated the results with glee.

Keys	Clients	Domain
22.31 Trillion	544	uiuc.edu and 128.174
22.30 Trillion	687	gatech.edu and 130.207

Table 14. UIUC Ahead of Georgia Tech Again, May 10

Kees Cook at UIUC took a look at the graphs to see total keys tested and found that UIUC had tested roughly 350 trillion keys in total, while Georgia Tech had so far tested about 80 trillion keys. Giving credit where it was due, though, Cook noted how quickly Georgia Tech brought so much computing power online.

Getting that many clients cranked up in such a hurry was indeed an accomplishment. Those Georgia Tech guys knew just what to do. Georgia Tech student Jason Bennett relayed the story on the DESCHALL mailing list.

> Actually, there's a great story here. A few weeks ago, [Sports Illustrated] released a poll saying the University [sic] of Georgia has the best mascot. Well, the local paper decided to run its own little poll on its Web site. Unfortunately, they forgot to check for multiple votes. So, some Tech people go on the [local Georgia Tech] newsgroups and persuaded some others to run scripts and hit the server with tons of votes. We won the poll.
>
> Flash forward a week or so, and Vincent (I think) posts a message about needing more people on DES, since MIT and [another] school [were] beating us. Like moths to a flame!

Just before the UIUC and Georgia Tech battle for first started, Rocke Verser made an important announcement.

27

Overdrive

Tuesday, May 6, 3:05 P.M.
The Ohio State University, Columbus, Ohio

Part of building the world's fastest DES-cracking system was getting as many "nodes" as possible—computers running DESCHALL client software. Ohio State graduate students Guy Albertelli and Justin Dolske were making sure that people who wanted to participate would be able to do so, even if they didn't have equipment that was exactly mainstream.

Among the machines that Ohio State had were the sleek black computers produced by NeXT Computer, Inc. Apple founder Steve Jobs started NeXT after losing a boardroom showdown with then-CEO John Sculley in 1985. Despite never enjoying the huge commercial success of Jobs' other projects, including the Macintosh and the Apple II, NeXT machines could be found in universities and other institutions populated by early adopters of new technology around the country.

Albertelli produced a DESCHALL client for the black NeXT computers by porting Rocke Verser's key-testing software written in the C programming language. (See page 93 for a description.) Though the number of NeXT computers running in the U.S. and Canada was small by comparison to other platforms, they still numbered in the tens of thousands. Enlisting the aid of these machines would not prove difficult because even in the companies and universities that had the black NeXT computers, the machines had largely fallen into disuse, meaning that there would be little competition for their resources.

After building the NeXT DESCHALL client, Albertelli put the software through the testing and quality control process that Verser imposed on all developers. Once the tests were finished and Verser gave approval to release the client, Dolske put the client into the DESCHALL client distribution archive at Ohio State and I did the same at Megasoft Online. Albertelli drafted a quick message to the mailing list that the NeXT client would be available soon. Over the course of several hours, several dozen distinctive NeXT computers joined the hunt for the key.

From his own NeXT computer (which was now running the DESCHALL client), project participant Michael D. Stanfill thanked everyone who worked on the production of the NeXT client. He noted that while the aging NeXT machine was testing a modest 2.4 billion keys in its first day of operation, it was happily contributing what power it had. Stanfill's fast 64-bit computers from SGI were testing keys at a much faster rate, but he started to wonder about the promised bitslicing clients for his 64-bit machines. It had about a week since Kindred shared the results of his tests with the DESCHALL mailing list readers, so Stanfill asked for an update on when those clients would be released.

While the mailing list had been quiet on the topic of bitslicing for a week, the developers were thinking of little else as Friday approached. Throughout the week, Kindred's bitslicing work was being integrated with Verser's fast DES key testing method, and the other developers built, tested, rebuilt, and retested the client software under their care. On Friday morning, the new version of the client software made it through the last tests and Verser sent the whole set to Dolske and me for inclusion into the DESCHALL client distribution sites.

Friday, May 9, 3:21 P.M.
Loveland, Colorado

Users of high-end engineering workstations with 64-bit microprocessors such as DEC's Alpha and SGI's MIPS had long been using the slower, portable client implemented in the C programming language. Fast, expensive machines with that slower key-testing software simply could not test anywhere near the number of keys per second that the relatively slow and cheap machines running Verser's fast software could. High-end system users' frustration would soon end.

A grueling week of client software integration, building, and testing was about to come to an end, but to make the most of it, Verser had to get his message posted before the majority of participants in the Eastern time zone went home for the weekend—and it was already approaching 5:30 P.M. there.

After receiving word from Justin Dolske and me that the DESCHALL distribution sites were ready, Rocke Verser posted a message to DESCHALL's announcement mailing list.

In a message entitled, "Ultrafast 64-bit clients!" Verser wrote,

> If you've got a 64-bit Alpha or SGI—stop running the DESCHALL client that you've got right now. Run (don't walk!) over to the DESCHALL archive and download the NEW client for your platform. These new clients implement a derivative of the Biham's bitslice method for doing DES encryption, and they're blazing fast.

These new clients, using the bitslicing method (described on page 151), more than tripled the speed of some of the clients. Considering that many of the largest contributors of processing power were universities with the kind of high-end systems that used these 64-bit processors, we expected the overall DESCHALL key testing rate to increase dramatically as soon as the new clients got put into place.

	Keys per second (in thousands)		*Relative*
Machine	*Old*	*New*	*Speed*
SGI Onyx2 (194 MHz R10000)	829	1943	230%
SGI Indy (180 MHz R5000)	370	820	220%
AlphaStation 255/233	451	1297	290%
AlphaStation 600 5/333	916	2944	320%
DEC Alpha 3000/400	200	600	300%
SGI PowerChallenge2 (R8000, 90 MHz)	233	820	350%

Table 15. Performance of First DESCHALL Bitslice Clients

DESCHALL long had the fastest DES key testing software—and some of the clients had just gotten much, much faster.

The day after the release of 64-bit clients for SGI and DEC systems, a new 64-bit DESCHALL client was released for high-end machines made by HAL Computer Systems, a subsidiary of Fujitsu. HAL's newest servers used the new 64-bit UltraSPARC chip designed by Sun Microsystems. Not many HAL systems were in use, but the increase in speed was dramatic enough to make the release worthwhile. In the first few days of the new 64-bit clients' availability, they were downloaded several hundred times. Sun's own UltraSPARC machines were much more popular, but a significant technical problem prevented us from being able to release a 64-bit client for Sun's UltraSPARC machines: even though the hardware was 64-bit, the operating system that governed Sun's computer was still 32-bit.

What the release of all of these bitsliced clients can be understood by analogy. Computers move data through their various parts like cars moving through a highway system. If a 32-bit system is like a highway of 32 lanes and 64-bit system is like having a highway with 64 lanes, a 64-bit processor with a 32-bit operating system is akin to a highway that is wide enough for 64 lanes but has only been painted for 32. Despite the extra width of the road, the painted "instructions" for how to use the road keep the traffic from being able to use the extra capacity.

Sun's first 64-bit operating system, Solaris 2.6, was still in the beta testing stage at the time—although Sun customers who really wanted it could get a copy of that operating system, it was not recommended for production systems. Since DESCHALL software was designed specifically not to interfere with normal operation, we did not encourage participants to upgrade their operating system solely for the benefit of DESCHALL. Any "upgrade" to a beta version of the operating system was out of the question. Hence, the only 64-bit client we issued for UltraSPARC-based systems were for the relatively rare HAL systems; the popular Sun UltraSPARC systems would keep needing to run the regular, comparatively slow 32-bit SPARC client.

Traffic coming into the keyserver was also increasing significantly. The keyserver software had a means of managing its own traffic load: it issued fast clients larger blocks of keys to test than slower clients. By managing block sizes, the keyserver could give clients enough work to keep them busy for an approximate period of time—with the target being about two hours.

Despite this safeguard, the bitslicing clients that we had released were taking the largest blocks that the keyserver would hand out and burning through them in just twelve minutes. Logs were kept on the keyserver, showing which clients got which blocks and what response the clients sent back to the keyserver. The amount of log data was growing, as we had not only more clients connecting every day, but because the clients were faster and thus connecting more often.

***Sunday, May 11, 12:07* A.M.**
Loveland, Colorado

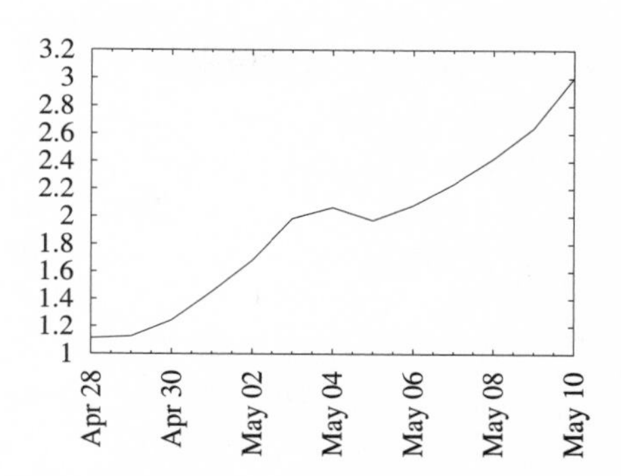

Fig. 7. DESCHALL Key Testing Rate (Billions per Second), April 28–May 10

As he did every night just after midnight, Rocke Verser looked at the log data for the previous day. As he was analyzing the data, Verser realized that two important thresholds had been crossed. Saturday's average key test rate ran at over 3 billion keys per second for the entire day, just a day after first achieving the rate of 2.5 billion keys per second. In addition to the sustained key testing speed, the total number of keys tested by DESCHALL thus far exceeded 3.6 *quadrillion*—five percent of the entire keyspace. As Verser was thinking about how far we had come, he started to notice some problems.

Packets of data containing messages from key-testing clients coming in through one of the two T2U gateways were malformed. This meant the packets were useless to the keyserver, and possibly indicative of a grave problem with some of the clients. Malformed messages would not only mean that the keys would need to be tested again, but that the additional load on the keyserver could push the system beyond its capacity and make it stop serving legitimate client requests.

Drawing upon years of training and experience as an engineer to stay focused on the problem, Verser observed the traffic closely, looking for what exactly was wrong with the malformed packets, comparing them to what they should look like, and considering what could introduce the problem.

At 1:54 A.M., the bad packets stopped hitting the keyserver. That was the good news. The bad news was that all of the other traffic to the keyserver had also stopped.

Verser tried to remain focused on the facts, not allowing himself to consider the consequences of reports for 3 billion keys per second being lost. Verser quickly performed a series of tests to identify where exactly the problem was located. The first test checked the connection between the keyserver and the ISP that connected it to the Internet. With that link responding properly, Verser looked at the next set of connections out toward the Internet.

Verser's ISP, Front Range Internet, Inc. (FRII) in Boulder, had redundant Internet connections, so if one of their Internet connections disappeared, the other should have picked up the slack. Verser's second test showed that the outage was between FRII and its connection to the Boulder Coop.

The next question was why FRII's other Internet feed, MCI, was not picking up the traffic. Verser remembered that the previous day, MCI had been experiencing network trouble, causing problems with packet loss between some of the clients and the keyserver. He thought that perhaps MCI was still having trouble while the FRII link to the Boulder Coop was down. The situation turned out to be much worse.

Routers are the special-purpose network computers that shepherd traffic around the Internet. Routers do this by keeping track of which networks they're connected to and which networks other routers are connected to. By keeping track of these "routes," routers know where to send an IP packet in order to get it closer to its destination. Routers direct packets from one network to another until they finally make it to their destination.

The Boulder Coop's routers should have detected the problem with the link and stopped advising other routers that they can deliver traffic to the network that fed the DESCHALL keyserver. Those routers, though, did not notice the problem, so packets destined for the keyserver continued to flow into the Boulder Coop, where they would sit, waiting to be sent over a connection that was down. After trying to send the packet to the next router for two minutes without success, the router would quit trying. Like a postcard sent to a nonexistent address, the UDP datagram with a message for the keyserver would be lost.

With the problem beyond his control, Verser proceeded with the routine he performed every night, working with the project statistics. After Verser logged into the system where he and Karl Runge did their statistical work, he decided to copy some of the data on to a writable CD-ROM.

Since the beginning of the project, Verser had been keeping copies of the keyserver log data by burning the logs to CD-ROM. Instead of using a different CD-ROM disc each time he made copies, Verser used a newer type of CD-ROM, the "multisession" discs—which would allow small parts of the disc to be used. Thus, instead of needing to use five discs for five different backups, Verser could use five “sessions” on the same disc to store those data as long as there was enough room.

Verser put one of his multisession discs into the statistics server. The CD-ROM drive spun up, and the server immediately stopped working. Verser could not get the system to eject the disc. Typing on the keyboard did no good: the system was completely frozen. Left with no option but to force a reboot by turning off the system’s power and turning it back on, Verser hit the power switch.

When Verser turned the server’s switch back into the “on” position, he watched the screen as the system went through its tests and began to boot. As the system attempted to bring its filesystems—formatted chunks of the internal hard drives—online, it reported that a problem had been found, and attempted to correct the problem. Verser blankly stared at the screen, hoping that the damage could be undone. When the first repair attempt failed, Verser finally began to worry. He took the system into its administrative mode and set to work on repairing the damaged filesystems.

After several more attempts, Verser got the structures repaired so the computer could bring the filesystems online. One final reboot, and the system came up properly.

Hoping to get something to work right before the night was out, Verser returned to the task of reporting of the previous day’s project statistics. Although the structure of the filesystems had been repaired, Runge’s statistics programs and directories had been destroyed. Fortunately, Verser still had copies of the data and the programs, but the statistical reports on the previous day’s work needed to be rerun.

Sunday, May 11, 6:11 A.M.
Front Range Internet Inc, Fort Collins, Colorado

As technicians finally restored connectivity between the Boulder Coop and FRII, the circuit came to life and a burst of data flew over the circuit, headed for the DESCHALL keyserver. In the nearly four and a half hours of downtime, reports for 43 trillion keys were lost, requiring that those keys to be retested later.

Still awake, though weary and frustrated from the night’s events, Verser noticed the flood of incoming traffic. He watched the traffic pour in to the keyserver from all of the clients requesting new blocks of keys to test and waited for the keyserver to catch up with the demand.

As the traffic normalized and returned to the regular, steady stream of reports and requests, Verser noticed another client beginning to flood the keyserver—sending the same message over and over. He made an entry to his firewall system to block the traffic from the misbehaving client.

With everything returning to normal by about 7:45 A.M., Verser sent private email to DESCHALL coordinators detailing what had happened throughout the night and then turned in for some much-needed and well-deserved sleep.

Karl Runge logged into the server for calculating statistics, restored his programs, and went to work, rerunning the calculations. By Monday, all was back to normal.

Monday, May 12, 10:52 P.M.
The Ohio State University, Columbus, Ohio

Looking over Sunday's statistics, Justin Dolske excitedly reported to the DESCHALL mailing list that while participants at the University of Illinois at Urbana-Champaign and Georgia Tech were fighting for first place, CMU snuck up from behind and took the number one position, as shown in Table 16.

Keys Tested	*Clients*	*Site*
20.06 Trillion	538	Carnegie-Mellon
20.05 Trillion	542	UIUC
19.09 Trillion	720	Georgia Tech

Table 16. Top Three DESCHALL Contributors, May 11

Organizations like the large universities could easily see their standings in the overall project statistics, which were reported by their domain name, such as uiuc.edu or gatech.edu. As the May 10 battle for first place between UIUC and Georgia Tech showed, not all participating systems had properly-working DNS. In such cases, the DESCHALL statistics was unable to report the contribution by domain name and would have to rely on abbreviated IP address in its reports. That turned out not to be the only anomaly in reporting.

James F. Eyrich, a network manager at Hoopeston Area Schools in Illinois, had been running DESCHALL clients on many of the machines in the school system. Like everyone else running the clients, Eyrich was interested in not only the overall project statistics, but the stats for systems under his control.

When Eyrich would go to Justin Dolske's *Graph-O-Matic* statistics site, instead of getting a pretty graph showing the domain's activity, he would get an error message, "no data found." On May 9, Eyrich finally sent email to Dolske to find out what the problem was.

Dolske searched through his raw logs and found an entry for Hoopeston, but it wound up with a different label from what would normally be expected. Usually, domains would be reported by the rightmost part of the domain (e.g., ohio-state.edu, frii.com, etc.) As it turned out, the domain was being handled as a special case, since it was hoopeston.k12.il.us. Searching for hoopeston.k12 would find the appropriate graph showing activity to date.

Armed with the new reports, Eyrich was prepared to show graphs of the project, and of the school's contribution to his boss, who had taken an interest in the DESCHALL project, and show him how the school's computers were contributing.

28

Distributed

Monday, May 5, 5:39 P.M.
MIT, Boston, Massachusetts

As the DESCHALL project started to acquire a large number of participants who had been working on the project for a while, the nature of the discussion on the mailing list began to change. In addition to the tactical discussions about how to keep the project growing, participants started to talk more philosophically about what implications projects like DESCHALL might have on the future of computing.

In his research at MIT, Nelson Minar was looking at how to build large, distributed computing grids. His interest in DESCHALL wasn't primarily driven by the cryptographic or public policy impact it could have. In his mind, no one who really understood cryptography had any illusions about DES being secure against a determined attacker, and burning off a thousand years' worth of Pentium Pro computations wasn't likely to change anyone's mind one way or the other.

On the other hand, there were serious questions about whether distributed computing could ever develop into a kind of global network of computing power that could be coordinated to attack large, complex problems.

The idea of having many clients work on problems that can be parallelized wasn't new either. Ever since the 1970s, when Xerox's Palo Alto Research Center (PARC) started working with "worm" programs that would copy themselves from one computer to another, researchers have thought about how to harness the power of many independent computers working together, cooperatively performing some computation that could not be done by one machine in a reasonable amount of time. By the mid 1990s, many universities were using the techniques on various computer networks for handling problems like rendering graphics.

Universities and research facilities weren't the only ones working in the area either. In 1995, Pixar amazed the world with its full-length feature film *Toy Story*. To render the photo-realistic images needed to create the film, Pixar used a network of Sun UltraSPARC systems working cooperatively as if they were one huge computer.

DESCHALL was different from these distributed computing efforts in a critical way, however. Rather than being a highly centralized effort to coordinate the effort of many computers all under the same ownership and administration, DESCHALL was only loosely coordinated. Like the projects that worked on RSA's factoring challenges since 1993, machines participating in DESCHALL had many different owners, with many different computing platforms involved, could start and stop their participation freely, and were communicating over a public network that had no single owner.

Some other distributed computing projects like DESCHALL were around. Adam L. Beberg and some others who had started on the RC5 Bovine project decided that it would be a good idea to have a single place to coordinate large-scale computations that could be tackled using distributed-computing methods. Besides the RSA Secret Key Challenge and Factoring Challenge contests, there were other groups harnessing computing power in the same way. One such project is SETI@HOME, a scientific experiment that uses Internet-connected computers in the Search for Extraterrestrial Intelligence (SETI). Participants run a free program that downloads and analyzes radio telescope data.

Another project is the Great Internet Mersenne Prime Search (GIMPS), which looks for prime numbers—those divisible only by themselves and one—in the special form of being one less than a power of two (that is, $2^P - 1$). For example, 10 is not prime because it is divisible by 2 and 5, in addition to 1 and itself. Five is a prime number, but it is not a Mersenne prime because it is not 1 less than a power of 2. Seven is a Mersenne prime, because not only is it prime, but it is 1 less than the eighth power of 2, i.e., $2^8 - 1 = 7$.

Seeing the success with earlier RC5 key cracks, factoring of RSA keys, and the progress of DESCHALL, RC5 Bovine coordinators had plenty of evidence to support their suspicion that distributed computing was ready to be taken beyond the realm of single organizations trying to harness the power of their own machines and into the age of the Internet tying together coordinating clients to perform large-scale computations. With this view of the future clearly in mind, Beberg registered the domain distributed.net and described the goal of his project as bringing together all the computing power possible to tackle a single task. From that domain, the RC5 Bovine group would undertake numerous large-scale computations after DESCHALL finished its work.

Meanwhile, the Coderpunks mailing list carried a debate on how to manage the integrity of such distributed computing projects and how to convince people to run the software needed to participate in those projects.

Minar watched the Coderpunks discussion but decided against participating in it directly. Instead, when the topic came up on the DESCHALL mailing list, he pointed out a key social aspect of the project that people looking at motivations from a purely theoretical standpoint were consistently missing. As his own experience with DESCHALL demonstrated, he had fun checking the project statistics every day, watching his university's progress, seeing how others ranked in comparison, and keeping track of just how many keys the project was testing per second overall.

Having some active role in a project that was testing over 2 billion keys per second and working with others had a simple social appeal that was proving critical for getting new participants and keeping them involved in the project after they started running the client software.

Saturday, May 10, 2:06 P.M.
MIT, Boston, Massachusetts

Even though our project was really moving, we had no time to rest on our laurels.

Since late April, Karl Runge performed statistical calculations to report not only overall progress, but to try to predict how our growth rates would continue. By plotting the number of keys that we had tested per day, Runge could predict what our growth was likely be in the future. By knowing the rate at which we could expect to increase our processing power and the amount of processing power needed to search the entire keyspace, Runge could predict when we would likely be finished and what our chances were of finding the key at various points between the present and total exhaustion of the keyspace.

Runge plotted the DESCHALL project's progress on a graph and sent it to the mailing list. In addition to the actual daily progress (the jittery line), Runge plotted a straight line showing linear growth, and a smoothly curved line showing exponential growth. Seeing the progress graphically was powerful, but when trying to predict where the project would be fifteen days in the future, there was plenty of room for debate. The difference between linear and exponential growth was not pronounced at this stage of the game, but it would make a big difference soon.

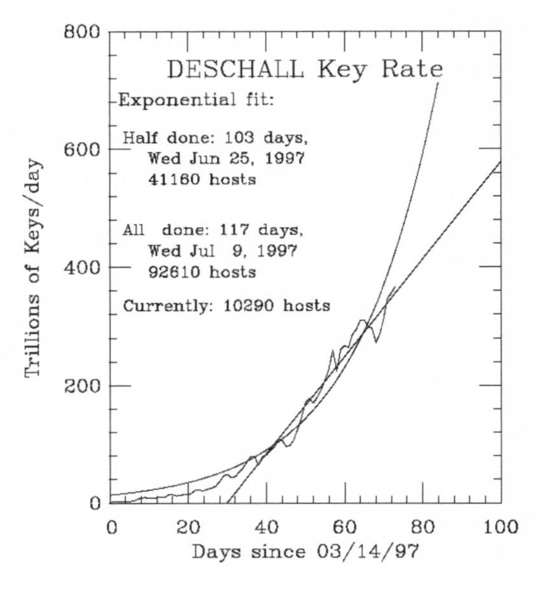

Fig. 8. DESCHALL Key Testing Rate, Compared to Statistical Curves

Looking at the same data, Dennis Okon at MIT did some analysis of his own. He built a model that showed how long it would take DESCHALL to get through the keyspace if the key testing rate followed an exponential growth curve, a steep linear curve that would plot to roughly sixty degrees, and a less aggressive linear curve that would plot to forty-five degrees. In that model, Okon showed that if exponential growth held, the DESCHALL project had a 50 percent chance of finding the right key by May 31 and a 100 percent chance of finding the key by June 8. If the linear growth remained aggressive, Okon's model showed a fifty percent chance of finding the key by July 10 and a 100 percent chance by August 8. Finally, moderate linear growth would get us to the halfway point on September 4 and through the entire keyspace on November 11.

When looking at data through May 10, Okon concluded that we actually reached the end of our exponential growth spurt and that we were instead following a path of aggressive linear growth, along the lines of Runge's linear growth prediction. After performing this analysis, Okon conceded that we really couldn't be sure of the impact of summer breaks on our computing power, although he thought that the school holidays could have a dramatic impact on his projections as the summer began.

Wednesday, May 14, 11:43 P.M.
The Ohio State University, Columbus, Ohio

Addressing the DESCHALL mailing list, Justin Dolske recounted the progress of the last few days. Sustaining a search rate of over 3 billion keys per second with 9600 computers running our client software, we crossed another threshold: a total of 4.4 quadrillion keys tested, amounting to just over six percent of the total DES keyspace.

While encouraging participants to continue to find new machines and new volunteers was important, DESCHALL developers and coordinators had to be much more than cheerleaders: we had to keep producing software that would make people feel motivated to stay involved.

Dolske's announcement included three new bitslice clients for 64-bit machines, namely Hewlett-Packard PA-RISC, HAL SPARC64, and SGI MIPS. We also had an improved version of the gateway software. Since the keyserver was keeping track of how clients were participating based on the IP addresses of the computers reporting keys tested, all participants using the gateways had their reports appear to come from the same computer as all other users of that gateway. The end result for the statistics was that all gateway users were participating anonymously—all gateway activity got lumped together and reported in aggregate. Some gateway users wanted to have their contributions reported uniquely so they could show their contributions just like other DESCHALL participants could. The new version of the U2T/T2U gateway software gave them that option.

Finally, a new Macintosh client was available. That Mac client was produced by the latest participant to join the ranks of DESCHALL developers, Andrew Meggs of Antennahead Industries, Inc. While the footers of his e-mail messages would carry titles like "Defender of the Universe," "Content Provider," and "Head 3D Superfreak," Meggs proved himself a first-rate programmer with an excellent mastery of the low-level workings of the PowerPC processor—the brains of the Macintosh. Putting his prowess to work less than a day after hearing about DESCHALL, Meggs worked with Verser to produce a hand-optimized version of the DESCHALL client for Macintosh systems. As a result, the same kinds of clever tricks employed to make the Verser's Intel clients so fast were now available to Macintosh users with PowerPC processors. That the complex and tedious work of hand-optimizing the client software for the PowerPC was finished less than two weeks after it started was nothing short of tremendous.

Unbeknownst to the project participants as a whole, Karl Runge had been compiling some great statistics on the DESCHALL client activity per platform. Seeing how many keys were being tested per platform became an important tool for us to see whether bringing the software to new platforms would likely be worth the effort. As we suspected, showing keys tested per platform became yet another way to encourage friendly competition among participants: the zealous Linux, Macintosh, and OS/2 users participating in the project could employ the statistics to convince their likeminded friends to run the DESCHALL client software on their own systems, so as to demonstrate for the world the popularity of their favorite computing platform.

Keys Tested	*Platform*
1322.516 trillion	Windows-Intel
446.102 trillion	Sun-Sparc
317.628 trillion	Linux-Intel
235.998 trillion	Unknown
200.052 trillion	MacOS/PPC
71.292 trillion	NetBSD-Intel
67.078 trillion	HP/UX-hppa
66.635 trillion	OS/2-Intel
56.358 trillion	AIX-rs6000
55.330 trillion	Irix-Mips
50.289 trillion	DEC-Alpha
43.722 trillion	SolarisX86-Intel
42.170 trillion	?
27.315 trillion	BSDI-Intel
24.766 trillion	DigitalUNIX
14.225 trillion	IRIX-Mips
1.838 trillion	MacOS/68k
1.088 trillion	Linux-Alpha
0.318 trillion	Linux-Sparc

Table 17. Platform Rankings from April 24 to May 12

Runge showed the number of keys tested per platform between April 24 and May 12 (Table 17), the one-week period from May 5 to May 12 (Table 18), and then just for May 12 (Table 19).

The per-platform rankings were a bit strange in that they showed minor differences in some versions of the same client for a platform. For example, the older clients for the SGI systems reported their platform as "Irix-Mips" while the newer clients were reported as "IRIX-Mips." The platform rankings included such different clients separately.

The Windows-Intel platform dominated the keysearching, due in part of the popularity of the platform and in part because of the sheer speed advantage of Verser's lightning code for the Intel processor. As clients for various other platforms were improved and as users of the less popular systems recruited like-minded advocates, client rankings changed slowly over time. Looking at the rankings for the past week, for example, showed that the Macintosh client for the PowerPC processor family (MacOS/PPC) became more important to the project overall.

Keys Tested	*Platform*
656.481 trillion	Windows-Intel
238.486 trillion	Sun-Sparc
156.061 trillion	Linux-Intel
120.894 trillion	MacOS/PPC
86.783 trillion	Unknown
40.444 trillion	?
35.889 trillion	NetBSD-Intel
31.935 trillion	OS/2-Intel
31.895 trillion	Irix-Mips
30.183 trillion	HP/UX-hppa
28.764 trillion	AIX-rs6000
24.766 trillion	DigitalUNIX
22.412 trillion	SolarisX86-Intel
19.512 trillion	DEC-Alpha
14.872 trillion	BSDI-Intel
13.926 trillion	IRIX-Mips
1.229 trillion	MacOS/68k
0.905 trillion	Linux-Alpha
0.205 trillion	Linux-Sparc

Table 18. Platform Rankings from May 5 to May 12

Unfortunately, there were several other issues with the platform rankings. The keyserver couldn't always tell which platform originated a report on keys tested. Two different clients fell into the category—

though we knew both were some variant of Unix—one was reported with a "?" and the other was reported as "Unknown." Whatever "?" was, it tested more keys than "Unknown" on May 12, though it was still behind for the week. Probably a number of the "?" clients were brought online and would need a few days before their total contribution would surpass that of "Unknown."

Keys Tested	*Platform*
115.0642 trillion	Windows-Intel
37.340 trillion	Sun-Sparc
27.084 trillion	Linux-Intel
19.229 trillion	MacOS/PPC
12.787 trillion	?
11.809 trillion	Unknown
6.585 trillion	NetBSD-Intel
5.437 trillion	AIX-rs6000
4.728 trillion	DigitalUNIX
4.374 trillion	OS/2-Intel
3.842 trillion	IRIX-Mips
3.658 trillion	HP/UX-hppa
3.440 trillion	Irix-Mips
2.606 trillion	SolarisX86-Intel
2.086 trillion	BSDI-Intel
1.618 trillion	DEC-Alpha
0.186 trillion	MacOS/68k
0.093 trillion	Linux-Alpha
0.064 trillion	Linux-Sparc

Table 19. Platform Rankings for May 12

29

An Obstacle

Saturday, May 16
Loveland, Colorado

Failures to route packets properly to and from the keyserver would not be the only obstacle for DESCHALL to overcome. While DESCHALL was setting DES key testing rate records during the weekend of May 9–11, participants were watching their machines' performance carefully. Demand for per-domain contribution reports from Justin Dolske's Web-based Graph-O-Matic soared far beyond the capacity of the machine to keep up: it simply didn't have enough processing power to issue the reports at the rate they were being requested.

Normally, such a hammering wouldn't have been a big problem. In this case, though, the Graph-O-Matic server turned out to be commissioned for several jobs—as was often the case in university environments—and all of them required memory and processing time. Besides trying to keep up with all of that Graph-O-Matic traffic, the machine was also serving network filesystems and routing traffic from one network to another. A machine like that crashing or otherwise being unable to keep up with the load was sure to get someone's attention.

Dolske turned off Graph-O-Matic and sent me a note to hasten the migration of the software to my site, which had a faster processor that would be better able to handle the onslaught of requests.

As I was working on the Graph-O-Matic migration, I received an e-mail message from Rocke Verser. Integrity of the DESCHALL project was critical, so coordinators typically encrypted and signed e-mail to other project coordinators with PGP, keeping the contents safe from prying eyes and assuring us that we received a true copy. After decrypting the message, I read its ominous contents: all developers should verify the security of our computer systems and the safety of the DESCHALL source code.

Verser's warning was prompted by the news that someone had broken into a DESCHALL developer's computer. The target of the attack was apparently DESCHALL project secrets, possibly including source code for the clients or detailed design specification. Another possible target could have been the DESCHALL project integrity, perhaps making malicious modifications to the DESCHALL client software that would undermine our project or harm the users who ran the client software. SolNET withstood an attack already, so we weren't really surprised to find ourselves in the crosshairs of some nitwit.

Analysis of the compromised machine showed that no confidential DESCHALL material was exposed. Apparently frustrated by the inability to get DESCHALL information, the perpetrator maliciously harmed the system, doing some damage to data unrelated to DESCHALL.

Verser asked each of us to ensure that DESCHALL confidential information was protected against unauthorized disclosure and damage. Although we had been taking precautions, such as encrypting source code and confidential e-mail among coordinators and developers, we needed to be vigilant and keep our guard up. DESCHALL project integrity was no longer an abstraction: someone, somewhere wanted to do our project, and possibly our users, harm.

Even though no one outside of the group of DESCHALL developers learned of the attack, there was widespread discussion of the risks involved with distributed computing projects like DESCHALL and SolNET that week. Much of the fuss was started by an article that appeared in issue 19.14 of the highly-regarded electronic newsletter *RISKS Digest*, published by distinguished security researcher Peter G. Neumann at SRI in Menlo Park, California. An article on client software integrity was submitted by Thomas König, a SolNET participant and graduate student at the University of Karlsruhe in Germany. Prompted to write about the issue of client integrity assurance by the recent attacks against SolNET from Finland, König stated that,

> You may remember RISKS-19.09, in which I discussed the risks in a network-wide attack on the RSA DES challenge: the Swedish group at http://www.des.sollentuna.se/ didn't give out

> its source, so the client could, in fact, do anything, such as crack a master EC-card key. The reason given was client integrity.
>
> Well, a month after this, the promised source code release has not happened. Instead, it appears that somebody disassembled part of the client, made a version that reported fake "done" blocks, and then sent these to the servers.
>
> Moral? Don't ever think that nobody can read compiled code. Don't try to run a cooperative effort like this in a closed development model.

Although DESCHALL developers agreed with one of König's conclusions (even compiled code can be read by someone with enough expertise), the criticism of the "closed development model" was significantly more controversial. Presumably König meant that distributed computing projects should give out the source code to their client software to prevent problems like SolNET's. König did not explain how giving out the source code would have prevented SolNET from being vulnerable to the attack from Finland.

Some DESCHALL participants would sometimes complain that we were putting DESCHALL at risk by keeping the source code and client development so heavily under wraps. We generally did not address these requests or criticisms. There was too much work to be done to get drawn into every possible philosophical debate. Fortunately, some DESCHALL participants were willing to address other, less sensitive projects and to take the initiative to get them done.

Dial-up users continued to contribute what they could to the effort, in spite of the tremeondous difficulty of keeping them in communication with the keyserver for long periods of time. These were some of our most dedicated volunteers, willing to put up with far more hassle to participate than more typical users who usually had access to faster and more reliable Internet connections.

We worried about more "typical" users, though, because we wanted them to join our effort. DESCHALL still needed more clients; on May 11, we had finished testing just five percent of the total keyspace. We needed more clients running and we needed things to be easy for would-be participants if we were going to get time on their machines.

Windows programmer Randy Weems joined the DESCHALL project in the first week of May. Just a little more than a week later, Weems posted an announcement of his own to the DESCHALL mailing list: the availability of a piece of software he wrote to help Windows users with dial-up Internet access manage their DESCHALL clients. His software was a graphical user interface for the standard DESCHALL client for Windows. He called his software DESGUI.

DESGUI ran as a small icon in the task bar, creating a hidden console window. It started the DESCHALL client and watched the output, but instead of just showing the DESCHALL client's text output to the user, DESGUI drew a graphical display of the client's progress and statistics. It also provided the ability to have the log files redirected to a file, so users who had software that read the DESCHALL client output could use DESGUI and whatever software they wrote at the same time.

Doing more than just providing cool features, the real motivation to use the software was to address the problems with dial-up networking in Windows. The most critical feature of DESGUI was its ability to connect to the Internet over a dial-up modem about one minute before the DESCHALL client would need to connect to the keyserver. By the time the client was ready to report its status and to ask for another block of keys to test, the system would be online. After the transaction with the keyserver, DESGUI would drop the connection. Operating this way, Windows users participating over a modem would be able to stay just as productive as the users of systems that were always connected to the Internet.

Because DESGUI was strictly a front-end for the DESCHALL client software, it was not actually testing any keys and was not actually a part of the DESCHALL client. This also meant that there was no need to coordinate the work needed to get the software tested and integrated into the core DESCHALL code.

Reaction was swift and clear. It was a big hit, a must-have for Windows users with dial-up connectivity to the Internet. There were a few bugs in the initial release, which were pretty quickly fixed, and DESGUI version 1.2 was released. Users reporting problems in the earlier releases reported that all was well with the new version.

Now that participants running Windows had DESCHALL client management software, Toronto-based system administrator Ken Chase asked how long a client had to report their findings before the keyserver would ignore the report. Chase naturally assumed that a report from a client about a block of keys that had not been assigned would be ignored because it was a false report. He also assumed that a report from a client about a block of keys that had been assigned a month earlier would also be ignored as a false report because anything with even the computing power of a wristwatch should be able to get through a block of keys in that period of time. But just how long did a client have to issue a report to the keyserver that would not be treated as a false report?

Chase asked the question because with DESGUI and the client management programs like the one that Benjamin Peterson at Notre Dame built and the one that he wrote himself, it seemed possible that one of those programs could stop a client from working for some reason, and keep the program stopped so long that the report would be ignored by the keyserver once it did finally finish. Based on some earlier commentary on the mailing list, Chase worked with the assumption that the timeout period was two hours and thought that the way his software worked might be problematic. Chase's software ran on Unix systems running DESCHALL clients. When a user logged into the machine, Chase's software would suspend the DESCHALL program's execution and wait for the user to finish with the computer. Once the user logged out, Chase's software would restart the DESCHALL client, letting it pick up right where it left off—although that could be hours later.

The concern was a valid one, but it turned out that the keyserver would not ignore such reports. Rocke Verser posted a reply to Chase's question and answered definitively: the keyserver was not presently timing out blocks at all. Of course, it knew which it had issued and when, and which had not been returned, so reissuing blocks could be done easily enough if they did turn out to be abandoned. With only six percent of the keyspace tested, though, there was no need to worry about retesting abandoned keys. Verser went on to say that in the beginning, the keyserver was automatically expiring such blocks it thought were abandoned, and it could do so again—but that the timeout period would be set for a day or two, not just a few hours.

The automated methods for key management were safe.

30

Export

Wednesday, May 14
Capitol Hill, Washington, D.C.

The House Judiciary Committee continued its work on liberating cryptography with the SAFE Act, which survived its first hurdle just two weeks earlier. At issue now was a provision that even many of the bill's supporters did not like: specification of new criminal penalties for the use of cryptography in committing a crime.

This provision prompted a letter to the bill's sponsor, Rep. Bob Goodlatte. Signed by representatives from twenty-six organizations (including American Civil Liberties Union, Americans for Tax Reform, Center for Democracy and Technology, and Internet Society), the letter argued that the provision tended to draw attention away from the criminal act itself and to cast "a shadow over a valuable technology that should not be criminalized."

The letter then discussed how such a provision would affect other more familiar and readily understood technology. "It may, for instance, be the case that a typewritten ransom note poses a more difficult challenge for forensic investigators than a handwritten note," wrote the authors. "But it would be a mistake to criminalize the use of a typewriter simply because it could make it more difficult to investigate crime in some circumstances."

During the two weeks since the letter was received, committee members had plenty of time to consider the argument. Representative William Dellahunt, a Democrat from Massachusetts, addressed the committee. Dellahunt proposed an amendment to the bill designed to limit the provision that caused concern among many of its supporters.

Following acceptance of Dellahunt's amendment, the committee passed the bill by voice vote, sending it along to the House International Relations Committee for consideration.

While Washington debated the future of cryptographic policy, corporations and individuals all over the world continued to buy cryptography products to protect themselves and their information against threats to confidentiality and integrity. Among the American companies unhappy to sit idle as foreign competitors continued to meet the worldwide demand for cryptographic protection was Sun Microsystems, based in Santa Clara, California.

Known as a pioneer in enterprise computing, Sun intended to address this important global market. The *Wall Street Journal* reported on May 19 that Sun would sell encryption software internationally through a Russian supplier, Elvis+Co., founded by former Soviet rocket scientists. Elvis+Co. would license the software from Sun and deliver it to customers overseas.

Sun officials said that they had not made the deal to subvert the cryptography export restrictions, but to deliver solutions that its international clients needed.

31

Getting Word Out

Friday, May 23, 8:50 A.M.
Megasoft Online, Columbus, Ohio

Looking for still more ways to publicize DESCHALL, I considered ways to reach people without relying on the Internet to get to them. Libraries, computer labs, and schools all seemed likely sources of potential participants, but we had no promotional material to hand out to people who might want to know more and then join the effort later. Hoping to encourage some creative thinking for flyer designs to fit this purpose, I wrote to the mailing list about my progress on developing different designs for DESCHALL fliers to be printed on standard U.S.-letter paper. I also added that I would post any good designs on my Web site, where they would be found easily.

David E. Eison of Georgia Tech responded to my note and mentioned he had a friend design a DESCHALL flyer, suitable for posting anywhere. The flyer included a critical visual—a pixelized version of Georgia Tech's mascot (a hornet named "Buzz")—which would not make sense for other sites. Nevertheless, he offered a few design tips that would prove useful for others making flyers of their own.

First, the text should actually be minimal, allowing for large and clear letters to be used. Second, the handout should include a Web address where interested people can go for more information; he also suggested having some tear-off tags along the bottom so people could take one instead of having to remember the address. Third, Eison suggested enumerating reasons for joining the effort. He observed that those reasons should include some serious incentives like cash reward and influence over government policy as well as some humorous ones like "giving your computer a hobby."

Later in the day, Brian Young, an Internet systems administrator at Oral Roberts University, thought about our need to recruit more participants. Looking over the number of keys tested per day in the past two weeks (Figure 9), he noted that our progress wasn't anywhere near the rapid pace that it had been earlier in the month. In fact, our rate of keys tested per second actually decreased over the past week. He wondered if the decrease in peak search rate might have something to do with university students being away and not running the clients on as many machines when they weren't at school.

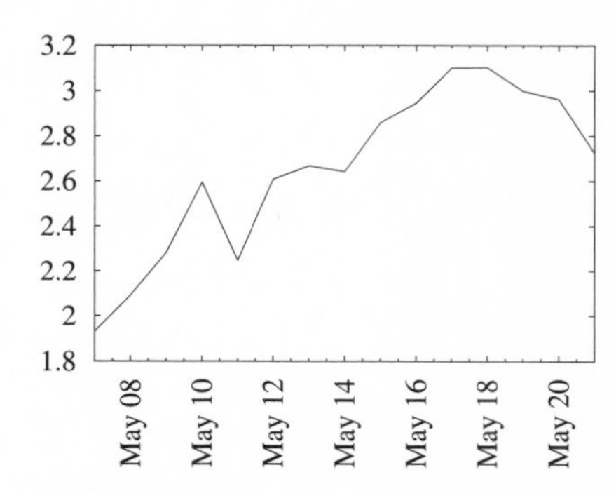

Fig. 9. DESCHALL Key Testing Rate (Billions per Second), May 7–21, 1997

Yes, we were slowing, but Young also noticed something that was more promising: we had passed the ten percent mark—more than 7.2 quadrillion keys had been tested by us so far. He suggested we consider a post-ten-percent press release in order to recruit new participants and inspire those who were already helping us.

Meanwhile, David Eison, who was also reviewing the most recent statistics, noticed that Georgia Tech's participation in DESCHALL had dropped off in the past few days. In response to Brian Young's questions about the general drop-off in key testing rates, Eison noted that Georgia Tech too had seen a decrease in its on-campus activity, although classes were not scheduled to end for another two weeks.

Wondering why this might be the case, he suggested that initial interest could be tapering off. Indeed, many people had, out of curiosity, been willing to download some clients and run them on their personal computers. Because their interest, Eison guessed, was more casual, these participants wouldn't install a DESCHALL client into their startup programs, and they wouldn't make any effort to keep the clients running. Thus, over time, as systems got rebooted, more and more clients run by people who only had a passing interest stopped working.

Since I was running the mailing lists, I saw some important data about interest in the project. In the week before May 23, I had seen more than 200 returned e-mail messages, almost exclusively from people at universities. Typically, mailing list administrators will see returned messages when subscribers abandon their addresses and the abandoned mailboxes become full, or when e-mail accounts used to read the mailing list are closed. We had plenty of subscribers in both categories. Either way, it looked like students had started to return home for the summer and were not reading their e-mail.

In other cases, people were running large numbers of clients without the knowledge or approval of their organization's management or system administration staff. As system administrators at different institutions became aware of the DESCHALL clients, they were deciding to ban the clients from their networks.

Something had to be done to ensure that DESCHALL clients would keep running. Eison and others he knew at Georgia Tech intended to organize a significant effort to keep DESCHALL clients running on campus and to spread the word about the project, hoping to counteract the effect of students going home for the summer. Eison hoped that others would do the same at their own universities.

While we were considering possible causes for decreases in activity, Eison noticed that no DESCHALL client was running at one of the buildings on the Georgia Tech campus where he knew they were installed. The key search rate for the project overall as well as Georgia Tech had been slowing in recent days, and this was something that would not help to reverse that trend.

Eison did some checking into the problem soon found the culprit. It turned out that one of the Sun machines had crashed while DESCHALL was running. Someone on the administrative staff there made an executive decision to terminate all DESCHALL participation in the building, believing that the DESCHALL program was responsible for the system's crash.

Unable to get any more details on what else was happening, Eison posted to the DESCHALL mailing list to find out if anyone else had experienced similar problems with their clients.

As Nelson Minar at MIT read David Eison's e-mail, he shook his head at the idea of a whole building full of computers unable to run DESCHALL because someone thought that a user process can crash a Sun machine. Incredible.

Operating systems like OS/2, Unix, and Windows are programs, Minar knew very well, but they're a little different from most other kinds of programs. The primary purpose of operating systems isn't for the sake of users, but for the sake of other programs. Operating systems provide an interface to the hardware for other programs, so that applications like word processors and web browsers don't need to know lots of intimate details about things like your video card, hard drive, and other components of your computer.

Another special property of operating systems is that they actually run other programs inside of them. So your word processor is actually a program that is running inside of another program—your operating system.

Programs do crash sometimes, when something happens and they reach some kind of situation that the designers never imaged. The "Y2K bug" that spooked computer users in 1998 and 1999 is a good example. If a program is designed to represent a year with two digits, that might be written as 97. Suppose that the program then has to perform a calculation that adds three to the year. Performing the simple addition of 3 to 97 will result in 100, which is obviously three digits long. If a program was written so that it could handle only two digits, putting a three-digit number into that part of the program's memory would cause a condition the programmer never foresaw, and the program would crash.

When people encounter something that they didn't expect, they might get confused, but they tend to recover quickly. For example, someone adding 3 to 97 would get 100, but would not try to put a three-digit number in place. A mistake like that by a person would be amusing, but that would be the extent of the "problem."

This difference between people and software is critical, and an excellent way to demonstrate how brittle most of the software built today really is. Obviously, programs that are written properly should not crash. Never crashing at all really isn't a practical goal for software developers, but we can eliminate enough problems in software that the *mean-time between failures* (MTBF) is so large that some other problem (like hardware failure) is likely to take the system down before our software crashes.

Operating systems tend to be especially robust software. If anything is supposed to be well-debugged and tested to be sure that MTBF is a "really long time" (months or years) it will be the operating system.

Most Unix-based operating systems had achieved quite a significant level of stability and correctness by 1997, with many being able to run for years without a reboot or a crash. IBM's OS/2 tended to be pretty good about stability at this time. Like Unix, OS/2 had *protected memory*, where programs could not write into parts of the computer's memory used by other programs.

To understand protected memory, let's reconsider the example of our program that added 3 to 97 and tried to store it in a place where only two digits were reserved. In an operating system without protected memory, the program would put the digits 100 into memory, starting at a given location. The 10 would fit, but then the next 0 would get written into the next section of memory—which might be used by another program. So now instead of having one program with a nonsensical value in memory, the program running next to it now was also compromised. The damaged portion of the program might not have been a number at all, but a letter, or a symbol. When that program tries to get that value back, it's going to get a 0 where something else used to be, possibly causing *that* program to crash as well.

Working under an operating system with protected memory, the buggy program that tried to put a three-digit value into a two-digit memory slot might crash, but it would not be able to make any other programs, or the operating system itself, crash.

Apple's MacOS could be wildly variant in stability, depending on which applications were running, since it didn't offer protected memory. Microsoft's Windows 3.1 was notorious for its instability, often requiring daily reboots, and sometimes more, frequently because an error in a program that the user was running would overwrite another part of memory, one used by the operating system itself. Once that happened, only restarting the system would correct the problem.

Sun's computers ran Solaris, a robust implementation of the Unix operating system—one with protected memory that could not be crashed because of any program that a user would run on the system.

Bearing all of this in mind, Minar posted the following response to the DESCHALL mailing list later in the afternoon:

> It's amazing how DESCHALL reveals what people don't understand about how computers work. Wearing out CPUs. Low priority processes pigging machines. And now, user processes crashing operating systems.

> The idea of DESCHALL crashing the operating system is ridiculous. If a user process as simple as DESCHALL can crash a Unix kernel [the core of the operating system] then that Unix is severely broken. Unix isn't Windows. [On the other hand], if the Unix system is truly that broken and I were [the system administrator] I would probably try to get people not to run programs on it, either.

As we considered how DESCHALL client software impacted the systems running it, one of our participants noticed a problem with how the client worked.

Friday, May 23, 9:43 A.M.
University of Oregon, Portland

Chris Schleicher decided to put some Sun machines running Solaris and some SGI machines running IRIX to work running DESCHALL clients during their idle moments. The Sun clients worked smoothly, but Schleicher noticed something a little different with the IRIX machines.

On several occasions, when someone tried to use the computer, the DESCHALL client didn't yield the processor the way that he expected, or the way that was happening on other Unix implementations. This was a problem because if DESCHALL took processing time away from users, Schleicher would not be able to run DESCHALL clients on those systems. If it was a problem with the client, he wouldn't be the only one who would have to pull the plug—other participants would have the same problem and probably have to quit running the client. Before we got to that point, we had to figure out why the client insisted on running when IRIX user programs had something to do.

The operating system's process scheduler is designed to work with the processor so that for each cycle (or, "tick") of the processor, the process scheduler will decide which program gets to be on the processor. Remember, a 200 MHz machine has some 200 million cycles per second; it's this fast moving among different programs that actually gives the user the illusion of running multiple programs at once. The process scheduler should be looking to see which programs want time on the processor and letting the DESCHALL client run only when no other program wants the processor.

Under the IRIX operating system, the DESCHALL client was still getting time on the processor, even when the user was running other programs that needed the system's processor. The end result was that users noticed a performance degradation when DESCHALL was running, which was something that we explicitly advertised would not happen.

Schleicher wanted to know whether the IRIX clients were faulty, or whether there was some problem with the way that the IRIX operating system was handling process scheduling. On the mailing list, we started to discuss how DESCHALL worked with process schedulers and to look into an answer for Schleicher's question.

The way that we were preventing the DESCHALL client from impacting system users was dependent on the "priority" value that we used when starting the software. This feature allows the process scheduler to understand when it has more than one program that wants the processor, which one should go first.

Most computer programs spend a lot of time in *idle loops*. That means that they're literally just sitting there, waiting for something (like input from the user) to happen. For example, a word processor will need to perform many functions: it will need to be able to accept input from the keyboard, figure out how to place the text, and draw the letters on the screen. Taxing as this might seem, the computer has much more processing power than what is needed for that particular job. The result is that other programs can run for tiny slice of time (the cycles, or ticks) between a user's keystrokes.

A problem could develop if, for example, the user were trying to use the word processor while sending a fax at the same time. If the fax program gets just as much time on the processor as the word processor, the user would wind up waiting for the fax program to yield its time so the word processor could receive and process all of the input the user gave it. To address this problem, modern operating systems have what is known as a *process scheduler*, a part of the operating system that will look at all of the programs that want to run and then decide which program will get to run for each cycle of the processor's time.

At this point it's important to note that not all of the programs running on a given computer have the same priority. Programs that users work with interactively need to be very responsive and thus cannot jockey with other programs for time on the processor.

Other kinds of programs do not have such tight time constraints. Performing a large scientific calculation, for example, can take days or weeks. Whether it takes an extra ten seconds or ten minutes will make no difference to the user. The user perception makes this kind of program dramatically different from a word processor, which could not make a user wait for a second before displaying a typed letter on screen.

The Unix family of operating systems has a basic facility for handling these issues. Each process runs at what is known as a *nice level.* That nice level is a number that represents priority on a scale between −20 (the highest priority) and 20 (the lowest priority). Programs have a nice level of 0 by default.

Nice levels help divvy up the system's processing power appropriately, since it allows more interactive programs to run at a higher priority than things that can go about their business without much concern over whether the job is taking thirty or thirty-five hours.

In theory, very low priority processes would be things that you wouldn't want to have running at all unless absolutely nothing else would run. This was the kind of priority that we intended to attach to DESCHALL clients. In practice, this turned out to be a litle bit harder than we originally thought. During the course of the project, we started to discover some minor differences in the way that various operating systems—even different operating systems that came from the same set of source code before they diverged—would handle process scheduling.

Darrell Kindred at Carnegie Mellon University was running DESCHALL on a variety of machines—some of which were running IRIX—so he took particular interest in Schleicher's question. Kindred noticed that the other Unix systems were behaving as expected, so he started to zero in on the way that IRIX specifically handled process scheduling. What he saw is that IRIX offered a much greater level of control over just how to share system processor.

IRIX provided a new command—npri—which set a program's priority, not in terms of the standard Unix nice value, but in terms of IRIX's granular *pri* value.

After having done this investigation, Kindred decided to start DESCHALL clients with a *pri* value of 150, which he understood to mean that any other process asking for time on the processor would get it. This solution was a much easier way to run DESCHALL than the nice command would allow.

Kindred always liked to know what was *really* happening in the computer. After posting his observations about IRIX's handling of process priority, he started to look at other Unix systems around the lab he worked in. He set about performing a series of tests during which he would start programs with very low priority set by nice, and then would run another program that wanted a lot of time on the processor to see just how much time the operating system would grant to the low-priority process. (Among the systems that he went about testing were an older SunOS 4 system, the newer Solaris, known also as SunOS 5, Hewlett-Packard's HP-UX, IBM's Unix implementation called AIX, NetBSD, and Linux.)

Something especially interesting—and rather frustrating—was that AIX did not provide the granular control over processes that IRIX did, and neither did a nice of 19 set the process priority to the low level that one might expect. Some investigation led to the discovery that under AIX, it was possible to do better than what nice was allowing.

Four hours after starting to look at the problem, Kindred posted again to the DESCHALL mailing list, summarizing his findings, and including the source code to a new program that he had written for AIX, which he called verynice.

At last, we had definitive answers to how we could ensure that DESCHALL clients would not interfere with normal system operations, even on operating systems that had non-standard ways of setting system priority. Schleicher simply used npri, and AIX users could use Kindred's verynice program. Other Unix implementations correctly handled the scheduling—not giving DESCHALL any time unless nothing else needed the processor.

Friday, May 23 was a frustrating day overall as we tried to understand why the DESCHALL key testing rate had slowed, saw students abandon their mailboxes, had DESCHALL clients banned from entire university buildings, and tried to track down strange platform-specific process scheduling problems. But across the Atlantic, things were much, much worse.

SolNET had been keeping hard on our heels. As a project, they were testing over 2 billion keys per second, compared to our 3.6 billion. Besides having so many systems running its clients, SolNET had been improving the speed of its clients. Though never as fast as DESCHALL clients, SolNET clients were getting faster, thanks in part to the work that SolNET did on its own bitslicing clients.

While DESCHALL participants discussed process scheduling in minute detail, the coordinators of SolNET were managing a real crisis. Analysis revealed that some of the clients that were already released were not testing all of the keyspace that they had been assigned. After removing the problematic clients from the distribution points, Fredrik Lindgren got the job of alerting SolNET users to the problem.

"We've found an ugly bug in the 32-bit bitslice clients," wrote Lindgren to the SolNET mailing list. "This will require everybody to upgrade their clients since we have to shut out all broken 1.11 clients. Sorry for this screwup."

Saturday, May 24, 11:58 A.M.
MIT, Boston, Massachusetts

Nelson Minar sighed as he read Lindgren's message. None of the participants really wanted to see any of the projects suffer any serious setbacks, even if they were our competitors.

After looking over the SolNET project statistics page, he posted his reaction to the DESCHALL mailing list.

> Since the announcement [SolNET's] keyrate has dropped from their recent 2 billion keys per second to 738 million keys per second...
>
> It's worth taking a step back and learning from SolNET's recent mishap. Is anyone on this list actively following SolNET's development? What exactly was the bug? Can someone comment on their quality control versus ours?

Minar asked good questions. Something that Rocke Verser took very seriously was the process of client development, control over the process of integration of source code, and testing of the clients before their release. Not knowing SolNET's development process, though, we could only speculate what might have happened there.

Minar continued,

> The costs of a bug in this type of computation are very high.
>
> I don't want this or my earlier e-mail comparing keyrates to sound like I'm denegrating SolNET. They are doing an excellent job of cracking DES [keys]. Our efforts are compatible,

any competition between DESCHALL and SolNET should be entirely friendly.

SolNET should be commended for their organization and openness. They have great mailing list archives, including an open development list. They are giving out source code for their clients (albeit with a small piece missing—you can't actually use the client to run on their keyservers). Making inforomation available makes it much easier to understand what is going on when a problem is discovered.

Saturday, May 24, 2:26 P.M.
Sun Microsystems, San Diego, California

Even while DESCHALL participants puzzled over SolNET's setback, we received some welcome news. Someone at Sun Microsystems had come on board. For the past two days, software engineer John Falkenthal had been running around trying to get everything that he needed to participate in DESCHALL.

Since Sun used firewalls to limit the connectivity between its internal networks and the Internet, Falkenthal needed to get Justin Dolske's U2T software working inside of Sun to work through the firewall. Once he got the necessary software up and running, he started running DESCHALL clients on the Sun systems at Sun's San Diego engineering facility.

The results of his first day of recruiting were impressive. Sun appeared in the DESCHALL statistics as one of the top ten sites testing keys on May 23.

Word continued to spread inside of Sun, and with other Sun employees starting to help by running DESCHALL client software on their machines, Sun Microsystems started making headway against other sites participating in the effort.

A few days later, Nelson Minar would smile as he read over the previous days statistics. John Falkenthal's efforts inside of Sun seemed to be paying off. Sun reached fourth place in keys tested May 27, passing both MIT and Georgia Tech.

Monday, May 26
Loveland, Colorado

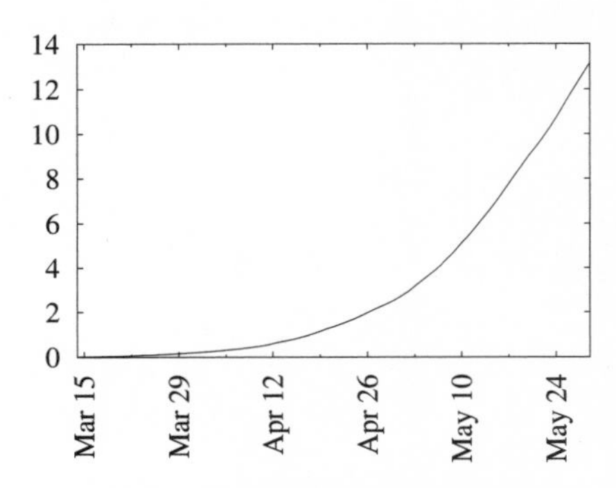

Fig. 10. Sum of DES Keys Tested, as a Percent of Total, March 14–May 29

Rocke Verser looked at the speed the DESCHALL project was searching the DES keyspace. It had taken us roughly two months to search one percent of the keyspace. We passed the two percent mark nine days later, the three percent mark six days after that, and the four percent mark in another four days. As we built momentum and raced through the keyspace at an ever-increasing rate, our "Estimated Time to Fifty Percent Completion" statistic became less useful. Sure, it would be a good figure to have if we wanted to show how long it would take us to find DES keys on average with the DESCHALL project, but we were looking for one key, and we wanted participants to know when to expect to find the one key we wanted before they became distracted or discouraged and dropped out of the race.

Verser's answer was a "Time of Completion" report, posted to the DESCHALL list for the first time on May 28, analyzing the previous day's activity. In the report, he noted that DESCHALL's previous "time to fifty percent completion" figure rapidly became a meaningless statistic, that a more accurate measure was the time remaining to searching fifty percent of the remaining keys. Additionally, instead of statistically assuming that DESCHALL was the only project working on the problem, Verser built a model that looked at both DESCHALL and SolNET's activity. Verser's first report was as follows.

> Given the "factual data" and the "assumptions" below, the "results" of this model are believed (not guaranteed) to be mathematically correct to within one day and to within two percentage points.

Factual data:

DESCHALL keyrate over last 24 hours	4.014 billion keys/sec.
SolNET keyrate over last 24 hours	1.738 billion keys/sec.
DESCHALL keyspace [tested]	12.193%
SolNET keyspace [tested]	7.849%

Assumptions:

- DESCHALL and SolNET are working independently.
- Each group's keyspace rate remains constant at the levels shown above.
- Nobody else on the planet is working on the RSA DES Challenge.

Results:

- 50% done: 54 days from now. (This is when $\frac{1}{2}$ of the total keyspace will have been searched by DESCHALL and/or SolNET.)
- Expected date of solution: 73 days from now. (There is a 50% chance the solution will be found prior to this date, and a 50% chance the solution will be found after this date. This is the date on which we should expect to have searched 50% of the remaining keyspace.)
- Probability that a DESCHALL client will find the key: 79%
- Probability that a SolNET client will find the key: 21%

Verser's new model provided a much more comprehensible view of a difficult problem inherently full of uncertainty. The more we could do to help regular computer users, members of the press, and average citizens understand what we were doing and what to expect, the better the chances that they would listen to our point and see the need for a new standard for data encryption. Verser's model was a big step in that direction.

32

Salvos in the Crypto Wars

Wednesday, May 21
Murray Hill, New Jersey

Bell Labs computer scientist Matt Blaze was well known for his work in cryptography. Neither was he a stranger to the place where cryptography and public policy met head-to-head. Among Blaze's work was a 1994 paper entitled, "Protocol Failure in the Escrowed Encryption Standard," in which he detailed a pair of attacks against the "Clipper Chip" cryptography products that were the centerpiece of the Clinton administration's key escrow system.

Blaze was updating the final version of a new report to his personal Web site, www.crypto.com. His co-authors included some of the most recognized and respected names in computing and security. A few of the names were Ross Anderson, professor of security engineering at the University of Cambridge; Steven M. Bellovin, AT&T Bell Labs distinguished member of technical staff and longtime contributor to Internet security; Whitfield Diffie, longtime critic of short cryptographic keys and coinventor of public-key cryptography; and Peter G. Neumann, distinguished computer security researcher and moderator of the popular *RISKS Digest*.

Entitled "The Risks of Key Recovery, Escrow, and Trusted Third-Party Encryption," the report documented the findings of the first thorough consideration of risks and implications of government-designed key-recovery systems by respected authorities on cryptography and security.[19] Among its observations came a pronouncement that would call into question the feasibility of implementing the Clinton administration's cryptographic policy:

> The deployment of a general key-recovery-based encryption infrastructure to meet law enforcement's stated requirements will result in substantial sacrifices in security and cost to the end user. Building a secure infrastructure of the breathtaking scale and complexity demanded by these requirements is far beyond the experience and current competency of the field.

For four years, the Clinton administration had made strenuous efforts to restrict the deployment and use of cryptography, largely in the name of making the Internet a safer place—by allowing governments the ability to police the Net. The first serious report considering the likely effect of such a policy was now out, and it argued that the policy would have exactly the opposite effect intended.

Reaction to the report was swift. The very day of its release, a statement was issued by Senator Patrick Leahy from Vermont. Leahy, the ranking Democratic member of the Senate Judiciary Committee, was working feverishly on legislation that would liberate cryptography from such tight government regulation. Aside from his own "Encrypted Communications Privacy Act," Leahy was chief co-sponsor of the Pro-CODE bill.

In response to the report issued by Blaze and other respected cryptographers, Leahy wrote:

> Last year the National Research Council concluded that aggressive promotion by the U.S. government of global key recovery encryption is not appropriate at this time. This new study by nine world-renowned cryptographers further shows the real-world problems with the government's proposal. It is even clearer now that the time for global key recovery encryption is still not right, and it may never be right. The U.S. government acts as though it doesn't understand the issue.
>
> Many of us fully expect that some users—maybe even many—will want and voluntarily choose to use key recovery encryption systems for some purposes. For example, no company wants to be left without a key to decode important business information stored in encrypted form on computer discs.
>
> The government apparently already is spending about $8 million on pilot projects to test key recovery systems, and that

> is just a drop in the bucket. According to the cryptographers' report "a global key recovery infrastructure can be expected to be extraordinarily complex and costly." As Congress examines the Administration's proposals for key recovery systems, we need to ask the questions about how much their proposals will cost the government, businesses and Internet users who want the strongest but cheapest security possible for their computer communications.
>
> Federal law enforcement officials contend that their objective is simple: easy, surreptitious access to both encrypted communications and encrypted stored data. The experts do not think this is so simple. The cryptographers' report observes: "We simply do not know how to build a secure key management infrastructure of this size, let alone operate one." When the experts say they do not know how to do it, we in Congress should think twice before legislating encryption commandments that may be impossible to afford and enforce.

Even while the rules for cryptography regulation were being debated, some companies were using the new Export Administration Regulation framework so their products could be used globally by American companies. After the government's investigation into Phil Zimmermann over the appearance of his Pretty Good Privacy (PGP) software on the Internet in 1996, Zimmermann went on to start a company to bring the system to market. PGP, Inc., based in San Mateo, California, announced on Wednesday, May 28 that the U.S. Department of Commerce had approved the export of its 128-bit encryption software to the overseas offices of the largest companies in the United States. The only restriction was that the offices were not located in any of the countries on the U.S. State Department's "T-7" list of terror-sponsoring nations, namely Cuba, Iran, Iraq, Libya, North Korea, Sudan, or Syria.

This made PGP the only U.S. company currently authorized to export strong encryption technology not requiring key recovery to foreign subsidiaries and branches of the largest American companies. Any other company wishing to export cryptography would have to limit their systems to the weak 40-bit systems like the one Ian Goldberg defeated in under four hours, or enter into an agreement with the government to develop a key recovery or key escrow system—and even then only up to 56-bit systems could be released.

More than half of the Fortune 100 companies already used PGP domestically to secure their corporate data and communications. PGP and 128-bit systems like it were preferred by people who wanted to keep information confidential for long periods of time because its strength against brute-force attacks was simply staggering. Recall that just two months earlier, Deputy Director of NSA William P. Crowell testified to Congress (see chapter 7) that all of the computers in the world in 1997 working on breaking a 128-bit PGP message using brute force would need 12 million times the age of the universe to find the key. With a 56-bit key length, DES was rapidly moving toward falling to brute force attack with only a few thousand computers over a period of a few months.

With the announcement, Robert H. Kohn, PGP vice president and general counsel said that the company "still oppose[d] export controls on cryptographic software, but this license is a major step toward meeting the global security needs of American companies."

PGP's announcement was not the only cryptographic news of the week. Weekly newspaper *Business First* in Columbus, Ohio, carried a "Tech Watch" column, where ongoing developments in technology would be reported to business readers. John Frees' Tech Watch column carried the headline, "Scientist questions standard for encryption technology." That column explored the important technical, political, and business issues at stake, and even published the address for the DESCHALL Web site.

The business press was not alone in starting to look at the issue of cryptographic policy. Business leaders themselves were also deeply concerned. On June 4, major computer industry heads openly called upon President Clinton to drop efforts to regulate encryption technology. In an open letter to the president, Microsoft's Bill Gates, along with the heads of Adobe Systems, Autodesk, Bentley Systems, Compaq, Intel, SCO, Symantec, Claris, Digital Equipment, Lotus Development, Novell and Sybase said that U.S. competitiveness in electronic commerce was at stake in the debate.

"Network users must have confidence that their communications, whether personal letters, financial transactions, or sensitive business information, are secure and private," Gates and his colleagues wrote as members of the industry trade group, Business Software Alliance (BSA).

As the BSA letter began to circulate, FBI director Louis Freeh testified before the Senate Judiciary Committee. He said that Congress must give FBI "the capability to deal with current and future technology" by increasing the Bureau's electronic surveillance authority. Freeh's testimony described how unbreakable encryption would "allow drug lords, terrorists, and even gangs to communicate with impunity."

Freeh then outlined his support for the kinds of controls sought by the Clinton administration and opponents to the SAFE Act. He said that key recovery or escrow systems must be put in place—going so far as to argue that these must be required even for domestic use.

What Freeh did not address is how making "unbreakable encryption" illegal would stop drug lords, terrorists, and gangs from using it, or why Congress could expect that people already committing crimes would be inspired not to commit the crime of encrypting their communications.

In their letter to President Clinton, the computer executives said governments should not impose import or export controls on encryption products nor "attempt to force use of government-mandated key management infrastructures." This position was in agreement with the analysis of the risks of key escrow systems detailed two weeks earlier by Matt Blaze and his colleagues.

As more weighed in on the debate, the lines hardened: business and technology experts demanded free cryptography, while the government demanded the ability to read encrypted electronic messages.

33

New Competition

By June 1, our principal rival in the race to defeat DES had recovered from the bug in its client software. SolNET released new clients and within a week, participants had upgraded, resulting in a recovery to a key testing rate of nearly 2.2 billion keys per second. In the meantime, DESCHALL had increased to over 4.2 billion keys per second. But we were in for a surprise.

With the late-May appearance of Sun Microsystems on our list of sites contributing processing power, speculation was growing about the absence of several other large technology companies. Attention focused especially on one of Sun's biggest competitors, Silicon Graphics, Inc., known simply as SGI.

On June 3, Sun's John Falkenthal posted some details of a rumor that he heard involving SGI's attempts to find the DES key, apparently hoping that someone else might be able to fill in the details. In addition to having an internal project of its own, Falkenthal wrote, SGI's effort rumored to be ahead of DESCHALL. He didn't know anything more—how far ahead, when it had started, or its key testing rate.

If an effort were seriously underway at SGI, it could very well have been ahead of DESCHALL. As a premier developer of high-end computing equipment, SGI had tremendous computing power available on its in-house machines. Furthermore, SGI would have the talent needed to create heavily optimized clients for various models of SGI computers. Even in 1997, all of SGI's computers were 64-bit machines and would have had the potential for extremely fast clients. SGI clients could have been running for months at speeds that DESCHALL clients were just beginning to see, and it could have thousands of fast, high-end machines at work. We had nowhere near enough information to guess just how many machines they were using, just how fast their clients were, or how far along they were, but we could guess that SGI might really be the front-runner by a large margin that would be hard to close.

Like any company with hot technology, SGI was usually pretty anxious to show off what it could do, which is why its absence in a contest like RSA Data Security Inc.'s was so noticeable. A secret internal SGI team working on the contest was certainly plausible and would explain why neither DESCHALL nor SolNET received support from inside of SGI.

The most interesting detail of the rumor was that although SGI was ahead of DESCHALL in total keys tested, we were actually sustaining a higher key testing rate. So even if the rumor turned out to be true, SGI might not be the front-runner forever, and DESCHALL could overtake them.

Not long after Falkenthal's message was posted, a follow-up was sent to the DESCHALL mailing list. A DESCHALL participant using the name "Stunt Borg" posted that he asked a friend of his at SGI about the rumored effort and soon after received a firsthand account of SGI's private DES Challenge project. "There is quite a campaign internally," reported the SGI employee whose name was removed from the message. E-mail notifications were being widely distributed, Web pages were put up internally, and even a pop-up window announcing the project was sent to all internal users. A survey was even issued to determine who was contributing and how much spare processing power they had available.

That night, Ken Chase in Toronto received a curious e-mail message that contained a table showing the top contributors to SGI's effort. Although the numbers leaked would not allow for highly meaningful comparison of the effort's total progress, the "identifier" column of the "top 20 participants" table did show participation from a broad cross-section of SGI. Participants were at SGI proper and Cray, the famed supercomputer maker, which SGI had purchased some time earlier. (Interestingly, Cray was also the manufacturer of the "$30 million supercomputer" that Robert S. Litt from the Department of Justice said would need over a year to crack a DES message.) Looking with more detail, Chase saw that even within these business units, various offices were contributing cycles, as were departments including security, marketing, and software development projects.

Chase forwarded the message to the DESCHALL mailing list in the late evening. Several hours later, Chase somewhat sheepishly composed another message to the DESCHALL mailing list. It turned out that he had received two messages about SGI's effort—the one that he had forwarded and one with more details on the effort's status. He overlooked the message containing the details. (Chase's error was a textbook demonstration of how making major changes in a computing environment tends to lead to strange changes in behavior. Chase had just switched to the Dvorak keyboard layout, which positioned its keys differently from the QWERTY system that is most common. The switch increased the amount of effort needed to use the keyboard, since Chase was still training his fingers to go to the correct positions. Consequently, Chase was typing less than he might otherwise, and would therefore avoiding opening messages that looked like they might require a response unless he was ready to put forth the effort to type with the Dvorak keyboard.)

Realizing what he had skipped over, Chase typed a brief explanation and forwarded the other message from inside of SGI to the DESCHALL mailing list.

While not detailing the scope or progress of the effort, the message did show that there was a fair bit of activity; it named the organizers and gave compelling evidence that the organizers had made every effort to make participation within SGI as easy as possible. Much of what they had on their internal Web site advertising the need for clients was what we had on our web site, that is, an explanation of the contest, the benefits involved, and how to participate.

On June 5, Rocke Verser privately reported to Justin Dolske and me that he had sent e-mail to the coordinators of SGI's DES Challenge effort. The leaks that we had been seeing included things like the organizers' contact information, and the name of the internal lists used for coordinating the effort. So why not try to contact them?

If there was truth to rumors about the extent to which SGI was working through the keyspace, it could have quite a significant impact on Verser's "Expected Date of Completion" reports, and he wanted to be sure that he was using the most complete information. A model that reflected SGI data in addition to DESCHALL and SolNET would be more accurate.

Now brought into the open, the coordinators of the SGI project were forthcoming and revealed two important sets of statistics.

The first statistic was the amount of keyspace tested. SGI had already finished just over 19 percent of the total 56-bit keyspace, notable because that was the greatest amount of keyspace covered by any of the known efforts. Second place went to DESCHALL at about sixteen percent, and third went to SolNET at just under ten percent.

The second statistic was the approximate key search rate, weighing in at about 2.8 billion keys per second. That came in a bit ahead of the 2.1 billion keys per second tested by SolNET, but significantly behind the 4 billion keys per second being tested by DESCHALL.

Overall, the figures showed SGI was in the lead, but DESCHALL was gaining fast and would soon pull ahead.

34

Netlag

The Internet's growing pains continued into June of 1997. One symptom was that some people could not access our keyserver from time to time due to small network failures. The DESCHALL mailing list became peppered with messages from participants who were having trouble reaching the server and who wanted to know whether others were having similar problems.

This leads us to a second "ailment" that afflicted the Internet's health at this time. Operations that used to take just a few seconds could take minutes, or even hours, as evidenced by lag time on the mailing list. Some people started to notice that there were delays in getting their messages posted to the DESCHALL list, and some users complained that it often took hours for the mail server to deliver their messages.

As it turned out, the mail server was working just fine, but several sites had difficulty with their domain name service (DNS) servers, the directory service that converted people-friendly names like frii.com to computer-friendly network address numbers like 216.17.184.30 and back again. Partly as an anti-spam measure, mail servers were just starting to refuse to talk to each other without being able to look up names and numbers in the DNS correctly. As a result, sites with DNS problems could sometimes sit idle for hours before our mail server could communicate with theirs and let the mail exchange properly.

The combination of these two problems could make for a frustrating experience. Someone thinks that the keyserver might be down and then posts a message to the list asking whether the problem is local or more widespread. Thanks to the sluggishness of different e-mail servers, an hour or two might pass before anyone received the concerned user's message. Meanwhile, whatever problem prompted the message in the first place—either a keyserver problem or an overload of traffic on some ISP—had been resolved.

Naturally, people getting the "Keyserver down?" messages would see that they could easily connect to the keyserver and quickly reply that all was working as expected. So not only was the original query rendered irrelevant by its delay, but it tended to prompt a flood of unnecessary responses.

The Internet basically worked and our project never would have been possible without it, but four or five years of exponential growth took a toll in the form of quality of service. Normally, no one would notice delays in e-mail delivery, but working on a project like DESCHALL, where everyone was tied together by way of the Internet and often trading time-sensitive information, would very quickly give one a heightened sense of how cumbersome the current Internet infrastructure still was.

35

Terminal Velocity

Thursday, June 5, 9:37 P.M.
The Ohio State University, Columbus

Our project's key testing rate depended on two factors: fast clients and many participants running those clients. DESCHALL's clients were already faster than any other project's, so it seemed to make sense that at some point, we were going to reach a limit on just how fast we could make the software work reliably. At that point, the only way to increase the project's overall key testing rate would be to get the key-testing software running on more computers.

But having software that was faster than the programs our competitors were using wasn't good enough. Justin Dolske posted an announcement to the DESCHALL mailing list on June 5: new clients were available, and they ran faster than ever.

Leading up to the June 5 release, Darrell Kindred improved the performance of his bitslice software again and then adapted it so the technique would also work on 32-bit platforms. Consequently, many clients experienced a performance increase, ranging from five percent to eighty-two percent.

The eighty-two percent performance increase went to the users of Sun UltraSPARC systems. Other Sun systems, based on older varieties of the SPARC processor had a forty-five percent increase. Clients for HP systems increased fifty-seven percent, Alpha clients increased twenty-five to forty percent, and AIX clients had a twenty percent performance boost. The same software release also included a new Macintosh client—complete with a slick graphical interface and a nice five to ten percent performance increase.

Almost immediately, the reports started to flow in: people were seeing dramatic speed improvements.

Several hours after Dolske's announcement, Rocke Verser was looking at the project statistics back in Loveland. Having updated his model for determining the expected date for a solution to the RSA DES Challenge, he put June 5's figures into the model.

What made the new model different was that it factored in *three* projects, rather than only two—adding SGI to the contenders for the solution. Using the data available on SGI's effort, Verser showed the key testing rates for all three projects.

Recent DESCHALL keyrate:	4.125 billion keys per second
Recent SolNet keyrate:	2.125 billion keys per second
Recent SGI keyrate:	2.890 billion keys per second

Table 20. June 5 Key Testing Rate Comparisons

In addition, Verser reported how much of the keyspace each project tested up to that point. For many participants, the report would be the first time that they saw how far ahead SGI was.

DESCHALL keyspace complete:	16.503%
Solnet keyspace complete:	10.090%
SGI keyspace complete:	19.573%

Table 21. June 5 Keyspace Completed Comparison

Working with the same fundamental assumptions as in the previous model—that each project was working independently, that their key testing rates remained constant, and that nobody else was working on the contest—some predictions could be made with Verser's statistical model. In particular, we could expect to find the right key in fifty-six days.

On Saturday, June 7, Karl Runge looked over the statistics he generated from DESCHALL keyserver logs. As he realized that he was looking at a report for another record day for DESCHALL, he smiled. The previous day's key search rate worked out to 4.4 billion keys per second.

Probability that a DESCHALL client will find the key:	51%
Probability that a SolNet client will find the key:	18%
Probability that a SGI client will find the key:	31%

Table 22. June 5 Probability of Success Comparison

Later on the same day in Reston, Virginia, Erik Fitchner, a Unix system administrator quietly contributing his machines' spare cycles to DESCHALL was paying close attention to the number of keys that his machines were testing. His 166 MHz Alpha machine started out testing 200,000 keys per second, and the first bitslice client for that platform brought performance up to 540,000 keys per second. He couldn't believe his eyes as the latest client reported that it was testing 940,000 keys per second.

Back in Toronto, Ken Chase was floored as he looked over the statistics for June 6. Like a race commentator watching the running order change as the leaders head into the home stretch, he excitedly reported the changes in the participants standings. Sun Microsystems had dropped to fifteenth place, Apple Computer jumped up to third place, and Penn State University dropped to fifth place. After rattling off the rest of the changes, Chase posted, "What's going on?"

At Sun Microsystems in San Diego, software engineer John Falkenthal read Chase's report. Determined to have Sun regain its position in the standings, he started up the new ultrafast client for the UltraSPARC processor on the latest, fastest, hottest machine that Sun Microsystems made—the brand new "Starfire" (Enterprise 10000) server, which had sixty-four processors.

That one Starfire server reported that it was processing 120 *million* keys per second—the same rate that the entire DESCHALL project sustained just about nine weeks earlier.

During the next week, I kept close tabs on our statistics. As client improvements started to stack up, we were testing keys at a phenomenal rate that just kept increasing. Back in April, when the project got going in earnest, we had used press releases to get the word out. After that time, we missed the opportunity to issue releases on our progress when we had tested ten percent and twenty percent of the keyspace. With twenty-five percent rapidly approaching and Rocke Verser's "expected date of solution" calculation going from hundreds of years down to weeks, it was now time to start thinking in less abstract terms about what to do when we found the key. Specifically, we had to be prepared to explain a lot of things to reporters and to help them to understand why this DES Challenge contest should matter to people who didn't know anything about cryptography.

Although the DESCHALL project coordinators all recognized the importance of handling the media correctly, we didn't get to begin any serious planning for when we found the key—a more pressing matter was before us.

Friday, June 13, 3:31 P.M.
Loveland, Colorado

About a week after releasing the most blazing fast clients imaginable, Rocke Verser composed a new announcement. The UltraSPARC users, long in a strange position because their 64-bit hardware was running a 32-bit operating system (Solaris 2.5), were about to find out that Darrell Kindred had performed a Herculean technical task.

Like a traffic engineer who could find a way to maintain 64 lanes of traffic on a 32-lane highway without causing any collisions, Kindred found a way to make UltraSPARC processors perform 64-bit operations reliably even though the operating system could only keep track of 32 bits at a time.

In addition, Verser found a way to improve the efficiency of the DES algorithm's "S-box" implementations on the UltraSPARC processor by ordering the events carefully and using extended logical instructions provided on the UltraSPARC processor.

The result was another increase in speed—of ninety-seven percent over the UltraSPARC clients released just the week before. An UltraSPARC system always running the latest client would have gone from testing 669,000 keys per second two weeks ago to testing 1.22 million keys per second last week, and then up to 2.4 million keys per second with this new client.

On the following afternoon, John Falkenthal at Sun fired up the latest client on one of the 64-processor Starfire systems. He gleefully reported to the DESCHALL mailing list that the *one machine* was now testing 239 million keys per second—more than the entire DESCHALL project was before April 5.

Putting the new client on a dual-processor UltraSPARC desktop (with 336 MHz processors), Falkenthal watched it sustain a testing rate of 9.716 million keys per second—roughly the processing power of the project in the first week of March.

On Sunday, June 15, Virginia Tech computing engineering junior Scott Coleman examined his university's statistics through Graph-O-Matic. With no small amount of pride, Coleman pointed out the tremendous leap in their standings was due to a single machine running the new client.

In Seattle, Washington, Scott McDermott, a system administrator for the King County Library System, was also running the UltraSPARC clients. Confirming Coleman's observations, he noted,

> I've only got three machines running: a pair of [older SPARC systems] and an Ultra 1. We were placed in the bottom half of the 300s [near 400th place in keys tested] Friday. I switched the Ultra to the new client and we jumped up to 213 on Saturday. I was most impressed!

Also on Sunday, at Oklahoma State University in Stillwater, Colin L. Hildinger noticed that DESCHALL wasn't alone in making advancements. Hildinger went over to the SolNET Web site and noticed that they had just released a new, faster Windows client, boasting speeds close to the speed of the Pentium clients offered by DESCHALL. While SolNET struggled to keep pace with DESCHALL on the Pentium processors, the latest round of DESCHALL clients for 64-bit processors further widened the gap in key testing speed.

DESCHALL clients were already running faster than many of us would have thought possible at the beginning of the project. We had used so many tricks to get the key-testing clients to run so quickly that we simply could not rely on improving overall DESCHALL testing speed by increasing key-testing software speed further. We needed more clients running, which meant finding more participants, but that weekend, we weren't going to worry about it. We just let the new superfast clients do their work while we tried to unwind from a stressful week of getting new software out the door.

DESCHALL developers could use the rest. Not only had we been working to get the new UltraSPARC clients released, but we spent extra time going through keyserver and gateway logs to ensure that the key reports we were getting were legitimate. The alarm that drove us to perform the extra work came up just as we were preparing to release the ultrafast 64-bit UltraSPARC client for Solaris, when Justin Dolske checked SolNET's mailing list to see how they were doing.

Thursday, June 12, 10:39 P.M.
The Ohio State University, Columbus

Dolske took interest in a thread of discussion on the SolNET mailing list about a surge of activity. Michael Fahlbusch, a SolNET participant from the Technische Universität München in Munich noticed a huge spike in SolNET clients from the Czech Republic. Excited to see so many clients coming online at once, he asked how the Czechs managed it, hoping to find the method they used to get more people participating in the SolNET project.

Jan Pazdziora, a network administrator and graduate student from the Czech Republic was also participating in the SolNET project. Pazdziora looked at the SolNET statistics for the Czech Republic after seeing Fahlbusch's message and came to an unnerving conclusion: another set of hostile clients had been launched to make false key testing reports to the SolNET key server.

Dolske wondered if the attackers attempted to get the DESCHALL clients before deciding to cause problems for SolNET. If that were the case, they would have had to attempt to download a DESCHALL client, so Dolske examined his client download logs to see whether we had been getting any requests from the Czech Republic. There was one attempt—which had (correctly) been rejected—three weeks earlier, but it wasn't from a university, much less the same one from which the attack against SolNET was launched.

Dolske next wrote to Rocke Verser, Karl Runge, and me pointing out that SolNET had been attacked again, and named the Czech network from which the attacker hit SolNET.

After getting Dolske's note, I checked the log for the client download site under my control, and did not find evidence that our clients were sought by SolNET's attackers.

In addition to looking for evidence of attempted downloads, Dolske also recommended that we do some additional analysis of the key-server's logs and check for any blocks that did not have the expected reports of "half-matches" (see page 93) that legitimate DESCHALL clients would make while testing keys. Blocks that had unusual patterns or amounts of half-match reports could be indicative of a false-report attack, so the extra step of detection would aid in recovering from an attack should it occur.

We really didn't have a lot of time to deal with things like extra analysis, but integrity of the project was of the utmost importance, so when there was a particular need, we'd make an attempt to deal with it. In the end, we don't know why attackers chose SolNET and left DESCHALL alone. We speculated at the time that the attackers sending false reports to the keyservers were engaging in the least interesting and most blatantly annoying kind of attack against the keyservers—just for a cheap thrill.

Because DESCHALL never published its source code or protocol specification, an attacker would have had to get the client and observe how it works, which messages it knew how to send, and how exactly it would format those messages. The relative difficulty of that reverse engineering work combined with the difficulty of actually getting the clients if not based in the U.S. or Canada might have been what saved DESCHALL from dealing with the attacks it did. Secrecy is not a sustainable strategy, but it might well have kept us from being the sort of "low-hanging fruit" that thrill seekers tend to target.

On the other hand, when DESCHALL did manage to get in the attackers' cross-hairs, the attack tended to be much more severe than false client reports, as the mid-May attack against a DESCHALL developer demonstrated.

36

Duct Tape

When asked, Internet engineers—particularly when living through the exponential growth in demand of 1997—will claim that much of the global computing infrastructure is held together by duct tape and bailing wire. Users rarely saw problems like delays in the mailing list when something behind the scenes broke; more often, participants would not notice or would only observe some strange side effects. During the DESCHALL project, we had plenty of occasions to work around some part of the infrastructure that decided to break for one reason or another.

Scott M. Hinnrichs, a participant from California went to the DESCHALL client archive on Saturday, June 14, to get updated client software. When presented with the question, "Are you a citizen or national of the United States, a person who has been lawfully admitted for permanent residence in the United States under the Immigration and Naturalization Act, or a Canadian citizen," he clicked "Yes."

Next he was asked, "Do you agree not to export the DESCHALL client software in violation of the export control laws of the United States of America? Or, if you are a Canadian citizen, are you obtaining the DESCHALL client software for end-use in Canada by Canadian citizens, or return to the United States, in a manner permitted by Canadian law?" He clicked "Yes."

Finally, he was presented with, "Do you assert that you have answered all of these questions truthfully?" Again, he clicked "Yes," after which, he moved his mouse over to click the "Submit" button.

Then he waited. Never having experienced a problem downloading the client software before, Hinnrichs sat there drumming his fingers, figuring that the server was just overloaded and that it would eventually give him the link to the clients.

At last, the server answered. Instead of providing the list of clients available for download, it displayed a message: "Export status not determined." Our download software would show this message when it was unable to determine from which country the request came.

Thinking that perhaps his computer's name and IP address were not mapping correctly, Hinnrichs checked his site's DNS. His tests confirmed that DNS was working correctly in both directions: name-to-address and address-to-name. That ruled out his site as the source of the problem, so he sent a message off to the DESCHALL mailing list asking what might be happening.

After he and I exchanged a few messages, I looked at the logs for the client distribution server. While far from overloaded, there was a problem.

Our client archive was designed not just to ask people if they were in the U.S. or Canada, but to make an attempt to verify the likelihood of the claim. As described on page 145, the software did this by looking at the verified domain name of the requesting machine and then finding that domain's country of origin by querying the domain name registry service provided by InterNIC.

InterNIC was established in January 1993 as a cooperative partnership among AT&T, General Atomics, and Network Solutions, Inc. to develop and to operate various services for the Internet community with funding from the National Science Foundation. These services, collectively part of an Internet-wide network information center (NIC), included whois, a service that would report the names and organizations behind registered Internet domain names and networks. By looking at the registration information for the domains and networks of the users requesting the client software, we could reasonably verify that people claiming to be in the U.S. or Canada probably were.

At the time that Hinnrichs was experiencing problems, our distribution server was timing out on those whois queries. The InterNIC system simply never responded, and after two minutes, our server would give up, and our software would report that export status could not be determined. I tried some manual queries against the InterNIC's whois server, to no avail. The service was simply not working, and even the web site for the InterNIC registry service was down. There was nothing to do but wait, the last thing that anyone wants to do in a race.

On the same day that the InterNIC registry service was down, Rocke Verser's Internet service provider continued making network changes that had started the previous day. Details of the move were not made clear, as usually happened for customers with "dedicated" connections that were always on; Verser was only told that changes were in the works.

Keeping DESCHALL operating under such conditions was like trying to operate a business whose customers placed telephone orders at the rate of three per second while the main number was being changed and without the "owner" knowing what the new telephone number would be once "changes" were complete. Without details on the network changes, we could not work around any outages or other related problems. The best we could do was sit around and wait for things to break and hope that we could respond before any participants' clients needed to contact the keyserver.

Fortunately, Verser had the foresight to build a feature into the client to handle a situation like this long before the problem arose: the software would notice changes in the keyserver's address and use the new address in the event of a change. With nothing left to be done, Verser kept in close contact with his Internet service provider so he could update the keyserver's IP address as soon as it changed. Once that change was made, the feature in his client software would enable it to adapt. In the short term, changes like this presented a significant challenge—not unlike trying to write a document in a word processor at the same that someone else was upgrading the operating system. Even with these troubles, we did not worry about the long-term effects, since the Internet had time and again proved itself surprisingly resilient.

For applications like e-mail, an outage of several hours wasn't a big problem. Not that many people—particularly on Saturday afternoon—were going to be responding immediately to e-mail. DESCHALL was different, though. Our clients were testing tens of trillions of keys per hour, and our project would experience a huge setback if a significant portion of our clients couldn't reach the key server for a couple of hours.

At the close of what turned out otherwise to be a relatively uneventful weekend, at least in terms of efforts needed to keep systems running, Justin Dolske noticed another problem at about half past midnight, early on Monday morning. A network file system (NFS) server at Ohio State had died, taking down both of Dolske's DESCHALL T2U gateways that helped clients running behind firewalls communicate with the keyserver. Dolske dashed a message off to Verser, telling him of the urgent need to remove the Ohio State servers from the list of available T2U gateways. Verser quickly made the changes, leaving only my T2U gateway at Megasoft Online in New Jersey.

As soon as the change had been made, Dolske sent a note to the DESCHALL mailing list explaining what happened. He wrote:

> The two gateway systems based at [Ohio State University] will be dead for awhile, as their NFS server appears to have gone belly-up for the night. The third gateway server should not be affected by this.
>
> The DNS records for "deschall-gateway.verser.frii.com" have been updated to remove these [two] systems, so everything should be working smoothly shortly. Until your local DNS server gets the new records, gateway users will continue to see a number of "connection refused" or "timed out" messages.
>
> Unfortunately, these things happen.

The recovery time, measured by usual people-time was fast—only an hour or two, but at the rate we were going, that delay would cost DESCHALL a few trillion keys.

During the weekend of June 13–15, we had seen amazing growth, thanks to the release of the new, blindingly fast clients. At the beginning of the DESCHALL effort, the large corporate and university campuses with the sophisticated 64-bit machines were at a serious disadvantage (using only a few optimization techniques, showing mild performance gains) by comparison to the PC users who had highly optimized clients. (Since the smaller computers are much more common, performance gains made there would have a greater impact, so starting with the most common systems made the most sense. At the time that Verser was writing his fast DES routines, optimization methods for 64-bit systems like bitslicing had not yet been published.) The tables were now tipping heavily in the other direction: the more sophisticated machines were finally getting the clients that would take advantage of their capabilities.

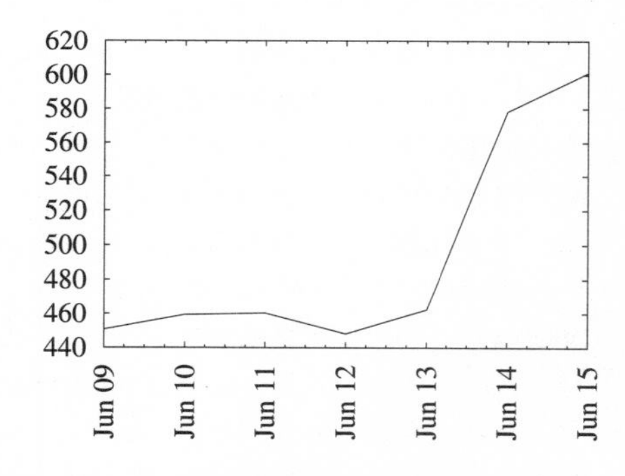

Fig. 11. DESCHALL Keys Tested Per Day (in Trillions), June 9–15

A look over the top site rankings and the performance per platform readily showed this to anyone who even glanced at those statistics.

As part of the fallout for that weekend's activity, people were once again asking whether people who were contributing relatively small amounts of processing power were going to drop out, with the likes of Sun, Carnegie Mellon University, and others putting extremely fast hardware with heavily optimized clients on the job.

After hashing out the issue once again, the conclusion was the same as it had always been: until such a point where we have so many clients that we are inhibiting the keyserver's ability to respond, each client mattered.

While the debate about the importance of small contributors continued on the DESCHALL mailing list, Verser composed a private message to several of us who had been discussing the issue on the list.

Addressing the issue of morale among the participants, Verser wrote that one week earlier, we had considered reporting how much of the total computational power each domain was contributing. In the end, though, we decided against it, because the statistics could easily be calculated by interested participants. Verser and others were also concerned that smaller organizations might give up, figuring that they were contributing "nothing" in the face of the horsepower that the likes of Sun Microsystems were throwing at the project.

Verser noted that daily statistics showed that roughly forty percent of the work was being done by small domains—those contributing less than one percent of the total processing power. Another ten percent was being done by even smaller domains—those contributing less than one-half percent of the total.

Having seen numerous universities and companies learn of DESCHALL and organize themselves to run the client software time and again, Verser wrote, "The nature of this project is that a given [organization] tends to rise rapidly to a peak, as the organization rapidly mobilizes most of their available computing power; and then (compared to others) the organization gradually declines." No matter how big an organization is, there's always one bigger, and no site can continue to double the work it was contributing every week, every other week, or even every month, as had been happening with the big campuses lately.

There was another reason for the smaller domains to continue participating. Ironically, as the "big fish" continued to join the project and to contribute more processing power, the odds of a "little fish" finding the key were actually increasing—on the basis of likelihood per processing cycle expended. At the beginning of the project, the chances of each key being tested was roughly one in 72 quadrillion (that's 72 thousand billion). By mid-June, the chances had improved to one in 55 quadrillion. As more keys got eliminated from the list of possible matches, the odds continued to improve—eventually reaching a one in one chance in the unlikely event that the correct key was the very last key scheduled to be tested.

Verser concluded that he was grateful for all contributions—each helped to bring the "expected date of completion" from some 200 years down to thirty-six days.

Tuesday, June 17, 8:29 P.M.
Northfield, Massachusetts

Seth D. Schoen was participating in DESCHALL in his senior year at Northfield Mount Hermon, a college-preparatory boarding school. After graduating earlier that month, he was able to spend more time on DESCHALL and began to think about the milestone that the project was rapidly approaching.

Because of working tirelessly week after week, we were going to cross the threshold of twenty-five percent of the total keyspace being tested in the next day or two.

Thinking that it had been a while since our last press release, or at least seeing the opportunity for a party, Schoen posted to the DESCHALL mailing list, asking "Is anybody planning either a party or a press release for the forthcoming twenty-five percent of keyspace completion?"

After articulating a few different angles that we could use in a press release to recruit more participants, Schoen concluded with, "If AP [Associated Press] picked up on DESCHALL ... I think the keyrate could get a pretty nice boost." It was a nice thought, and indeed every major news organization around the world would soon report on DESCHALL.

37

Showdown in the Senate

Tuesday, June 17
Capitol Hill, Washington, D.C.

The Secure Public Networks Act of 1997 was officially introduced by its sponsors, John McCain and Bob Kerrey in the Senate Commerce Committee. Offered as a compromise on the contentious cryptography policy issue, the bill had a striking resemblance to the draft legislation that the Clinton administration proposed in March—essentially requiring government access to keys used to encrypt data.

The other major cryptography legislation in the Senate—in many ways the opposite of the McCain-Kerrey bill (and the Clinton administration's proposals)—was the the Promotion of Commerce Online in the Digital Era, or Pro-CODE, Act, which was sponsored by Senators Conrad Burns and Patrick Leahy. Like the SAFE Act making its way through the House, Pro-CODE was intended to reform encryption policy, acknowledging that cryptography was already widely available, and allowing U.S. companies to participate in these international markets that were already being served by foreign competitors. The Senate Commerce Committee, chaired by Senator McCain, was set to vote on the Pro-CODE bill on Thursday, June 19.

The McCain-Kerrey bill tried to keep cryptography from being used by criminals against law enforcement officials but in reality did nothing to protect the privacy of Internet users or the security of their electronic transactions. In fact, it would require U.S. citizens to use key recovery systems approved by the federal government, require electronic commerce transactions to be conducted with government-approved key-recovery systems, allow the federal government access the keys needed to read encrypted messages without a court order, create new criminal penalties for people who used cryptography that was not on the government's list of approved systems, and codify the 56-bit key as the limit on products for export. All of this was proposed in the name of preventing criminals from using cryptography that could prevent government investigators from discovering and prosecuting them. Arguing that to allow strong cryptography is to be soft on crime makes for a tempting sound bite for political purposes, but hardly makes sense. Imagine seeing legislation proposed that requires citizens to leave the doors to their homes unlocked or to give a copy of their keys to the federal government, all for the purpose of making fighting crime easier. The notion is absurd.

Tuesday, June 17
Center for Democracy and Technology, Washington, D.C.

As soon as the text of the McCain-Kerrey bill was available, two Washington computer privacy advocacy groups went into action. The two groups—the Center for Democracy and Technology and Voters Telecommunications Watch—issued a press release and posted a call for action in the *Crypto-News* newsletter. The release characterized the new McCain-Kerrey bill as a false compromise which would do "nothing to protect the privacy and security of Internet users."

Instead, the new bill represented, the alert cautioned readers, "a full scale assault on your right to protect the privacy and confidentiality of your online communications."

In its description of the status of the situation, *Crypto-News* said:

> On Thursday, June 19, the Senate Commerce Committee is scheduled to hold a vote on S. 377, the Promotion of Commerce Online in the Digital Era (Pro-CODE) Act—an Internet-friendly encryption reform bill sponsored by Senators Burns (R-MT) and Leahy (D-VT).
>
> Senator McCain, the Commerce Committee Chairman, is expected to try and substitute his proposal for Pro-CODE—gutting the proposal and inserting provisions which would all but mandate guaranteed government access to your private communications.

> Please take a few moments to help protect your privacy and security in the Information Age by following the simple instructions below.

Following this call to action came a list of Senate Commerce Committee members. Readers were asked to call their senators, if they were on the list, and urge them to oppose the McCain-Kerrey bill. Readers whose senators were not on the list could go to a Web site and "Adopt a Legislator"—which would entail signing up to get targeted alerts whenever the adopted legislator would be nearing a vote on some Internet-related issue.

The people watching the legislative front of the Crypto Wars weren't the only ones were were foregoing sleep in the early hours of June 18, 1997.

38

"Strong Cryptography Makes the World a Safer Place"

Tuesday, June 17, 11:51 P.M.
Loveland, Colorado

A message contained in a UDP datagram made its way across the Internet at the speed of light. Originating in Salt Lake City, the datagram travelled up from a small local area network, to be handed off to a larger, wide-area network. Racing eastward, the datagram arrived in Loveland, Colorado at the DESCHALL key server.

The DESCHALL key server issued a message of its own to RSA's automated contest server, which almost immediately acknowledge our victory. Rocke Verser felt a jolt of excitement shoot through his body when he saw the "Key Found!" message flash across his computer screen.

Verser quickly wrote a message to DESCHALL coordinators and sent it on its way, encrypted with PGP, as was standard practice when writing about project details that were not yet public.

Wednesday, June 18, 2:12 A.M.
The Ohio State University, Columbus, Ohio

When Justin Dolske got the good news, he was ecstatic. As his motivation was to demonstrate the weakness of 56-bit keys, he would have been happy to hear that the key had been found by anyone, but being on the winning team made the victory all the more sweet.

Dolske laid out the steps necessary to notify the DESCHALL and SolNET mailing lists and to issue a press release. Leaving the message for the mailing lists to Verser, Dolske went straight to work on the draft of the official press release. His first draft would be circulated to DESCHALL coordinators at 2:56 A.M.

Wednesday, 8:04 A.M.
Megasoft Online, Freehold, New Jersey

I was spending the early part of the week in Freehold with the rest of my company's software development team. While I had been working in my hotel room into the wee hours of the morning, I was doing so completely without Internet connectivity and I had not checked my voice mail throughout the night.

When I arrived at the office on Wednesday morning, the receptionist handed me a handwritten message scrawled onto a torn-off piece of "greenbar" data processing paper.

The message was simple. "Call Rocke Verser. 'We found it!' "

I promptly sounded my "barbaric yawp" in the fine tradition of Walt Whitman and ran up the stairs and through the hallway leading to the software development offices. Finding a phone, I dialed Verser's number and waited for him to answer.

Rocke Verser had spent the entire night drafting the notices for the project participants, looking over drafts of the press release, and ensuring that the news was ready to be released to the world. Normally, this would be about the time that he'd be trying to catch a few hours' sleep, but that wasn't likely to happen today.

After a few congratulatory remarks, I put down the phone and got to work. We'd made history, and it was time for the world to know.

Coordination was critical to getting the story out right, because we needed to explain the project and its significance in a way that would make sense to readers and reporters who did not have a background in cryptography. The stories that we had seen so far were not particularly urgent in nature—mostly they had been local interest stories, where someone from the publication's readership was participating in a large-scale Internet-coordinated project. That kind of thing could be published any day of the week, and so any extra time needed to work through technical details could be taken. Now, the media would be in a race to get the story together and to get it out quickly—and that would mean less time for checking facts.

About an hour after the phone call with me, Verser had working drafts of the announcement for the mailing lists, the press release, and a data sheet with details of the facts of the project, intended to help reporters get the information they needed to get their stories together. RSA Data Security contacted Verser and worked with us to coordinate our press release with theirs.

Later in the day, when everything was in order, and the world was ready to know that it was a safer place because of cryptography, RSA and DESCHALL press releases were issued at the same time, and Verser proudly posted the message he had drafted earlier.

In an article entitled "WE FOUND IT!" Verser wrote to the DESCHALL mailing list:

> "Strong cryptography makes the world a safer place."
> That's the message RSA has been waiting for us to decipher. And we did it!
>
> The correct key (8558891AB0C851B6) was reported to RSA Data Security shortly before midnight last night (Mountain Time). RSA's automated server acknowledged our win!
>
> The winning computer is a Pentium 90 MHz, operated by iNetZ Corporation of Salt Lake City, Utah. Their employee, Michael K. Sanders, was the individual who was running the DESCHALL client.
>
> Congratulations, Michael. And congratulations to all who participated!

After acknowledging many individuals, he graciously turned to "our only public 'competitor,' SolNET," which he called "a class outfit." Verser continued:

> In a sense, the "win" belongs to all of us, who contributed CPU cycles and clients and ideas and innovations. We searched less than $\frac{1}{4}$ of the keyspace. Worldwide, over half of the keyspace was searched. A DESCHALL client may have found "the" key, but you [SolNET] deserve credit for helping to bring the "expected date of completion" significantly ahead.
>
> Your Web site gave us a goal to shoot for. A goal which we never met. Your clients had many features our users wished for. There is no shame in not finding the key. But I know the anguish you must feel after putting your hearts and souls into a project for three to four months and not being "the" winner.

> In my eyes, everyone who participated, whether working for the DESCHALL team or the SolNET team is a winner!

Elation followed on the DESCHALL mailing list, with congratulations and thanks flowing in from all around. Across the Atlantic, SolNET coordinator Lindgren Fredrik sent an announcement of his own to the SolNET mailing list. He wrote:

> The challenge is over ...
>
> ... and we "lost." On June 17 around midnight one of our competitors, DESCHALL, found the secret key and decrypted the secret text prepared by RSA Data Security Inc (http://www.rsa.com).
>
> The goal of the DES Challenge was to show that DES encryption could be cracked, and that better encryption is needed to keep data safe. In my opinion this goal has been accomplished.
>
> Although it's not that fun being a runner up, I must say that it has been an enjoyable couple of months running this effort. As much as it been a goal to show the weaknesses of 56-bit crypto, it's been very nice to be able to show the enormous amount of "surplus" computing power that is available on the Internet. Not to mention the warm and fuzzy feeling it gives me to think of everybody that's been working together towards the common goal of answering the challenge.
>
> The SolNET DES Team would like to thank everybody who has been participating and sharing their spare computing resources in our project. Without you none of this would have been possible.

Lindgren extended his thanks to many who contributed their resources and talents to run SolNET's keyservers, work on clients, and otherwise keep their effort running. He then pointed out the opportunities for additional work to be done, on the 56-bit RC5 Bovine effort, as well as the Great Internet Mersenne Prime Search.

Wednesday, June 18, 7:44 P.M.
The Ohio State University, Columbus, Ohio

Now it was time for the media to pick up the story and to weigh in. Late on Wednesday, June 18, Dolske was one of the first to be contacted, and answered questions for someone from "the Internet video show State of the Net." None of the coordinators had heard the program before, and we had no idea of its audience size. Dolske wasn't even sure that he was talking to a reporter, but the person with the questions was clearly interested in what had happened, so Dolske answered his questions.

After the conversation, he checked a search engine for "State of the Net," but couldn't find anything. Given the proliferation of "cyberculture" shows on television, electronic magazines, and streaming video, one could never really be sure. But it was someone interested, and it just might have been some of the first press the key-breaking received.

Verser meanwhile had spoken with reporters from Channel 2 in Salt Lake City, ZDNet, MSNBC, and the Chronicle of Higher Education. Obviously, the press releases were having their effect and the stories were being written. Some of these articles also included parts of interviews with Mike Sanders, whose machine found the key, and RSA Data Security officials.

RSA Data Security Inc., issued a press release of its own, at the same time as ours. In that release, RSA president Jim Bizdos was quoted, tying together the debate before Congress and the DES Challenge.

> RSA congratulates the DESCHALL team for their achievement in cracking the 56-bit DES message," said Jim Bidzos, president of RSA. "This demonstrates that a determined group using easily available desktop computers can crack DES-encrypted messages, making short 56-bit key lengths and unscaleable algorithms unacceptable as national standards for use in commercial applications.
>
> "This event dramatically highlights the fatal flaws in the most recent administration proposal, Bill S.909, 'The Secure Public Networks Act of 1997,' introduced by Senator John McCain (R-AZ) and Senator Bob Kerrey (D-NE). This bill, if passed, would severely hamper U.S. industry by limiting export to the 56-bit DES standard."

We would discover that yet another press release had been issued, without our knowledge, much less coordination. Sameer Parekh, an enterprising user of cryptography ran a company he started called C2Net Software, Inc. to bring products with strong cryptography to the markets. Never one to miss a media opportunity, Parekh sprang into action upon seeing that the DES Challenge had been won. Quickly he placed a call to iNetZ Corporation, where the 90 MHz Pentium machine run by Michael Sanders had found the right key. He got Jon Gay, a vice president at iNetZ to agree to a quote, wherein he hoped that the demonstration would cause users to demand strong cryptography in their products—"such as the 128-bit security provided by C2Net's Stronghold product, rather than the weak 56-bit ciphers used in many other platforms."

Parekh also got a quote from the respected cryptographer Ian Goldberg, recently of 40-bit Challenge fame. Goldberg's remark, "This effort emphasizes that security systems based on 56-bit DES or 'export-quality' cryptography are out-of-date, and should be phased out," was buried in an alarmist press release entitled, "Hackers Smash U.S. Government Encryption Standard." C2Net's press release failed to cite anything authoritative from either RSA or the DESCHALL coordinators, pointing instead to its own Web site that gave no additional information on the contest. The C2Net statement was entirely devoid of useful content about RSA's DES Challenge or the project that answered it; its entire purpose was to use DESCHALL's win as a platform from which to tell the world, in Parekh's words, "We refuse to sell weak products that might provide a false sense of security." Members of the press who saw Parekh's blatantly opportunistic commercial received it with some skepticism, some of which would unfortunately carry over into reporting into the facts of the DES Challenge.

DESCHALL coordinators granted many interviews that day, unaware of the C2Net press release. Had we known about the C2Net release, we could have taken the opportunity to put the matter into more balanced perspective than to suggest that the standard itself had been broken by "hackers." Reporters' deadlines finally came and the calls died down as the articles started to get written. As the reports started to make their way around the world, DESCHALL's coordinators got some well-deserved rest.

Thursday, June 19, 8:02 A.M.
Megasoft Online, Freehold, New Jersey

I was pleased to see the *Wall Street Journal* article on DESCHALL. A well-written article by Don Clark covered the contest and its impact, stuck to the story, and remained technically accurate. Many other reporters called on Thursday, following up with their own stories after seeing the early coverage of the news.

On Friday, the largest wave of media coverage came, and as DESCHALL participants saw the coverage, they posted their observations on the articles to the DESCHALL mailing list. Nelson Minar at MIT noted that the CNN article covering our work was subtitled, "But it took four months." Most media coverage had roughly the same flavor.

MSNBC's article managed to botch the story pretty badly, going so far as to assert that the entire keyspace had been tested as opposed to the one-quarter of the keyspace that actually had been tested. The *Money Daily* article carried the basic premise that our success was alarming, but readers didn't need to tear up their ATM card right away.

DESCHALL project did manage to get the attention of the mainstream media at a critical moment—as the capability to break messages encrypted with the standard came into the hands of even modestly funded groups of people and as the future of public policy was being debated. The success of the RSA DES Challenge would ultimately come not from what the media would say immediately, but whether we succeeded in "killing single DES," as Peter Trei wrote to the Cypherpunks on October 1, 1996.

39

Aftermath

Cracking a message encrypted with DES was a watershed event in the history of cryptography because we, private-sector cryptographers, participated in a large-scale demonstration of distributed computing to make our point. We knew that DES, the sitting standard for data encryption for twenty years, was vulnerable to brute-force attacks. We knew that finding a key wouldn't require a thirty-million dollar supercomputer and more than a year's time. So we quit estimating what it would take and just did it.

The contest wasn't just about cryptography in 1997. Cryptography's future was also at stake: we knew that long-term public policy was being debated by lawmakers in Washington under the influence of information specifically released to support the Clinton administration's legislative agenda.

We knew that the data encryption standard needed to be replaced, but no one would listen to us when we presented them with calculations. People would not listen unless we actually broke a message encrypted with the same system that was protecting sensitive information like their financial and medical records. So that's what we did. And then the world wanted to hear all about it and what to do about it.

Thursday, June 19
Gundaker Realtors, St. Louis, Missouri

Systems and security administrator Stuart Stock, who wrote the "DESCHALL Linux Bootdisk Mini-HOWTO," had been a participant for most of the project's duration. His efforts, and those of many people like him, got the project access to many computing cycles—easily twelve hours daily and two whole days weekly—that would have been otherwise unused.

Concerned that management might not have sanctioned the effort, Stock requested that his contribution be identified as an "anonymous" site in our statistical reporting—a request which we happily granted.

On the morning of June 19, Stock found himself answering some questions from the head of the company, who had seen the article in the morning's copy of The *Wall Street Journal* entitled "Group Cracks Financial-Data Encryption Code." The article got the head of the company thinking about the importance of strong cryptography in ensuring the safety of financial transactions and electronic communications.

Feeling more confident that he was making a connection and getting a sympathetic audience, Stock revealed to his boss that their company had been involved in the effort, finishing twenty-fourth in terms of contributed processing power.

After learning that Stock's method of contributing processing power had not interfered with business operations in any way, his boss relaxed.

Stock was satisfied with his contribution. The head of his company simply had no idea that cryptography was something he needed to consider. Like many people, he just assumed that things were "safe." Thanks to our project and the subsequent publicity, he was asking good questions and even being shown how to protect himself with Phil Zimmerman's Pretty Good Privacy cryptography software.

Since its beginning, Netscape had produced two versions of its software: one for domestic U.S. use and one for international use. The international use products were limited to 40-bit key strength, while the domestic versions used 128 bits.

On June 24, less than a week after our success in the DESCHALL project, Netscape finally was able to release its products with strong cryptography for export to the outside world with the permission of the U.S. Department of Commerce. Instead of having to fill out an online affidavit and go through verification that your system was based in the United States, users from all over the world could simply download the strong-cryptography version of the Netscape browser.

In addition, Netscape banking customers overseas could buy Netscape's server products with strong cryptography enabled.

"The ability to export our products with strong encryption enables Netscape to provide its customers worldwide with client and server software that can improve the security of their information and applications," said Taher Elgamal, chief scientist at Netscape in the press release announcing the change. "This approval is another example of Netscape's leadership in the privacy and security arenas and is especially important due to the recent breaking of 56-bit DES by the DESCHALL group last week."

On the same day, Microsoft announced that it got the same approval that Netscape did. In consequence, Microsoft would build 128-bit cryptography into its Internet Explorer 4.0, Money 98, and Internet Information Server products.

On June 27, Senate Majority Leader Trent Lott made a speech on the floor of the Senate, addressing the cryptography debate and the Commerce Committee's consideration of the Pro-CODE and McCain-Kerrey bills earlier in the month. As feared by proponents of unfettered cryptography, the McCain-Kerrey bill passed by voice vote in committee with very few changes, essentially gutting Pro-CODE and leaving the McCain-Kerrey Secure Public Networks Act as the main cryptography bill before the senate.

Senator Lott said,

> Mr. President, the demand for strong information security will not abate. Individuals, industry, and governments need the best information security technology to protect their information. The Administration's policy and the McCain-Kerrey bill allow export of 56-bit encryption, with key recovery requirements. How secure is 56-bit encryption? That question was answered the day before the Senate Commerce Committee acted. Responding to a challenge, a secret message encoded with 56-bit encryption was decoded in a brute force supercomputing effort known as the "DESCHALL Effort." The message that was decoded said "Strong cryptography makes the world a safer place."
>
> Now that 56-bit encryption has been cracked by individuals working together over the Internet, information protected by that technology is vulnerable. The need to allow stronger security to protect information is more acute than ever.

Conrad Burns, the Senator for Montana who was co-sponsor of the Pro-CODE legislation followed Senator Lott's address. Having heard law enforcement's concerns about child pornographers using cryptography that could circumvent investigators' ability to intercept suspects' online messages all throughout the debate, Senator Burns made an interesting observation. He said:

> As I sat through the markup last week, it occurred to me that we had allowed the issue of encryption to be framed as the issue of child pornography or gambling. I want to be sure that all parties understand that the reform of encryption security standards is not related to these issues.
>
> I have often said that encryption is simply like putting a stamp on an envelope rather than sending a postcard because you don't want others to read your mail. Encryption is simply about people protecting their private information, about companies and governments protecting their information, from medical records to tax returns to intellectual property from unauthorized access. Hackers, espionage agents, and those just wanting to cause mischief must be restrained from access to private information over the Internet.
>
> When used correctly, encryption can enable citizens in remote locations to have access to the same information, the same technology, the same quality of health care, that citizens of our largest cities have. Perhaps most importantly, it is about ensuring that American companies have the tools they need to continue to develop and provide the leading technology in the global marketplace. Without this leadership, our national security and sovereignty will surely be threatened.

40

Staying the Course

Wednesday, June 18
Chicago, Illinois

Adam L. Beberg and some like-minded volunteers had been working on creating a central site for Internet computing projects. Among the projects that caught their interest was the next of the RSA Secret Key Challenges, 56-bit RC5, often abbreviated as RC5-56.[20]

Beberg's distributed.net had started on the RC5-56 contest more than a month earlier (see page 201) but did not actively recruit from among the DES Challenge participants.

Happy to see the DES challenge solved, Beberg changed gears, openly and actively inviting veterans of the DES Challenge contest to his RC5-56 effort. "This time, we're all on the same team. DESCHALL, SolNET, [and] even SGI is invited," he wrote in his invitation posted to the DESCHALL mailing list.

Like Ian Goldberg (who defeated RC5-40 in three and a half hours), Germano Caronni (who defeated RC5-48 in 313 hours), and Rocke Verser's DESCHALL (which defeated DES in 140 days), the distributed.net group searched for a secret key needed to unlock an encrypted message.

On October 20, 1997, 265 days after RSA announced the contest, the distributed.net team located the secret key needed to read the contest message: "It's time to move to a longer key length."

RSA Data Security announced additional key-searching contests at its annual conference on January 13, 1998. The "DES Challenge II" was a pair of contests, just like the DES Challenge that DESCHALL answered the year before—with an important difference. The amount of the cash prize varied, depending on the amount of time needed to crack the message: if the winner found the key in one quarter (or less) of the time needed by the previous winner, the prize would be $10,000. A $5000 prize would go to a winner finding the key in up to half of the time of the previous winner; a $1000 prize would go to the winner finding the key in up to three-quarters of the time needed by the previous winner.

On February 24, 1998, distributed.net DES Challenge II (DES-II-1) project coordinator David McNett announced that DES had once again fallen to a brute-force search. The message "Many hands make light work" was decrypted—not in the 140 days that it took DESCHALL to find the key, but in a mere thirty-nine days. Especially interesting was the fact that rather than searching only one quarter of the keyspace, as DESCHALL had, the distributed.net DES-II-1 answer came only after search more than ninety percent of the keyspace.

On July 13, RSA launched the second DES Challenge II contest (DES-II-2). Again, distributed.net turned its attention to the contest. With the additional computing power that became available in the six months that had passed, and the fact that almost certainly less of the total keyspace would need to be searched, the previous record was certain to be beaten again.

Fifty-six hours after the start of the contest, DES-II-2 was solved, not by distributed.net but by the Electronic Frontier Foundation (EFF), a non-profit civil liberty advocacy group, in conjunction with Cryptography Research, a firm headed by cryptographer Paul Kocher.

With funding from EFF and the support of civil libertarian, EFF board member, and cypherpunk John Gilmore, Paul Kocher and his team at Cryptography Research designed and implemented "Deep Crack," a custom-built machine created for the specific purpose of cracking DES keys. Proving the assertions made by private-sector cryptographers true, Deep Crack showed that customized hardware—coming in at a cost of roughly $250,000—could crack cryptographic keys dramatically faster than any software.

Finally, in December 1998, RSA announced another contest to crack a DES message: DES Challenge III, to begin on January 18, 1999. The first to crack the message would receive a prize of $10,000 if doing the job was completed in under twenty-four hours, $5000 if it took under forty-eight hours, and $1000 if it took fifty-six hours. Anything longer would get no cash prize.

Twenty-two hours and fifteen minutes after the beginning of the contest, the message “See you in Rome ([at the] second AES Conference, March 22–23, 1999)” was extracted from the challenge ciphertext. The method was once again brute force, this time with distributed.net and Deep Crack working cooperatively and achieving a key search rate of 245 billion keys per second when the correct key was found.

The secret message in the DES Challenge III had special significance for cryptographers: the second AES Conference to be held in March 1999 was part of NIST’s effort to find a replacement for DES, which had reigned as the U.S. government standard for more than twenty years. The effort to define AES, the Advanced Encryption Standard, had been announced in the January 2, 1997 issue of The *Federal Register*. That article carried a note of particular interest. “It is NIST’s view that a multi-year transition period will be necessary to move toward any new encryption standard and that DES will continue to be of sufficient strength for many applications.”

Now more than two years after the announcement of the AES effort, it was clear that a multi-year period for the definition of a new standard would be needed. Also clear was the insufficiency of DES for any commercial or governmental application.

41

In Retrospect

Understanding an event's significance is usually pretty difficult at the time. Putting it into historical perspective and looking at how it influenced other events, though, can help a great deal. Since June 1997, there has been plenty of time to think about what we accomplished.

Could the Internet be the basis of a future computing platform—"the supercomputer for everyman" as Rocke Verser called it? Many people believe so.

Since the mid-1990s, and continuing through today, there are several kinds of projects that are attempting to harness this kind of computational power.

Other projects related to cryptography include key cracking projects, such as the distributed.net Bovine effort formed during the height of DESCHALL and started in earnest after the first fall of DES. That project has since solved RSA's first DES Challenge II and the DES Challenge III, as well as RSA's 56-bit and 64-bit bit RC5 Secret Key Challenges.

Finding large prime numbers is another example of a large computing project. The largest such project on the Internet is still running, the Great Internet Mersenne Prime Search (GIMPS), coordinated by George Woltman. (That project is being run from www.mersenne.org.)

We have shown time and again that the kind of computing power that can be harnessed using the Internet to coordinate many processors is phenomenal. Not all large computing problems are well-suited for this approach, but for a great many that are—problems that are actually made up of many small and independent problems—the possibilities are endless.

While the social and technical issues that DESCHALL and projects like it have addressed are of interest, ultimately, DESCHALL is a story about cryptography.

Even at the time, no one who understood cryptography and computing was surprised by what DESCHALL accomplished. That DES keys could be broken by brute force was understood from the beginning—even if the feasibility of such attacks was up for debate. We understand that exhaustive key search is an effective means of defeating any symmetric cryptosystem, save the Vernam Cipher, which is better known as the One-Time Pad.

Like the rest of security, cryptography is a tool that allows the defender to change the variables in the game against the attacker. At its most fundamental level, cryptography is simply a matter of economics. The whole idea is to make the target harder to reach for an attacker than the attacker thinks it's worth.

Make an attacker spend one million dollars to steal one million dollars, and you have taken away his economic incentive. The same fundamental principle is true even when the attacker's motivation is not money. Whatever it is that the attacker wants, if he has to spend too much of whatever he has to achieve it, he's better off simply following the policy that defines expected use and behavior.

We were able to demonstrate to the world what all of us already knew by calculation: 56-bit ciphers just aren't secure against dedicated attackers. Even with no better attack than brute force available, attackers without special equipment would be able to break the messages quickly enough to be worthwhile against information whose value extended beyond a few months. Subsequent breaks of DES messages demonstrated that the curve continued. At the end Deep Crack demonstrated that with a relatively modest initial investment (of, say, $250,000), a machine could be designed and implemented to break DES keys in a matter of hours.

Some might be inclined to argue that the cost of breaking DES keys at this point had become $250,000 and one day. I do not share this view. The amount of time needed to break DES-encrypted messages with such a system would indeed be one day, but $250,000 was not the cost for breaking the message—that was the cost for getting into the message-breaking business. Designing and building Deep Crack was a one-time expense.

Had EFF wanted, it could well start a key recovery business: by deploying the key-cracking system so that it would crack one message after another, continuously working around the clock and throughout the year, the entire cost of the hardware design and implementation could be covered in a single year by charging $685 to crack a DES message. The cost could drop further by processing enough volume to require a second Deep Crack system—the majority of the cost was in *design* (which would not need to be undertaken again), rather than in *implementation* (the only cost incurred in bringing a second Deep Crack online).

With this kind of startup fee and pricing schedule, even a small company could get into this business—as could any modestly-funded criminal or terrorist organization.

Considering how long medical records, credit card numbers, census data, and other kinds of information need to remain confidential, minimally attackers were shown to be a real threat to the security of this information. Funded attackers were barely slowed down by the defense of cryptosystems with 56-bit keys.

In reality, finding a replacement for DES was no more critical because of the RSA DES challenges. The security of DES was the same as it had always been and its susceptibility to brute-force attacks was in line with what we had predicted. But as a result of Peter Trei's October 1996 challenge to the Cypherpunks and RSA Data Security's support, thousands of cryptographers, programmers, civil libertarians, and hobbyists took the time to demonstrate for the public the critical need to heed our warnings.

At long last, an alternative to DES became a standard. On November 26, 2001, NIST Federal Information Processing Standard Publication 197, "The Advanced Encryption Standard" (AES) was published. The multi-year process of moving away from DES could at least begin. On July 26, 2004, NIST announced its proposal to withdraw DES as a standard altogether. In that announcement, NIST said simply, "DES is now vulnerable to key exhaustion using massive, parallel computations." The proposal's request for comments period ended on September 9, 2004. It would seem that in answer to Peter Trei's October 1, 1996 question, yes, we can kill single DES.

Instead being limited to 56-bit keys, we now have a standard in AES with variable key sizes available, providing as much as 256 bits of protection.

DES might have been replaced without the RSA DES Challenges—the process for replacement did start at NIST before DES fell to a brute force attack. On the other hand, NIST's AES announcement did come after RSA announced that it would launch the contests, and the failure of DES to withstand three public brute force attacks between 1997 and 1999 might have proved to be just what was needed to keep pressure on NIST to follow through with the standard.

What is less clear is whether cryptography would be free today without the DES Challenges. DESCHALL and its successors were often cited by lawmakers who kept efforts to repeal restrictions on cryptography alive in Congress; efforts of lawmakers to limit cryptography failed in 1997. Subsequent debate over cryptography continued, until the SAFE bill—reintroduced into congress yet again in 1999—began to pick up broad support. Even Senator John McCain, who had worked to defeat cryptography liberalization efforts in the Senate, became a believer in the virtues of free cryptography and supported SAFE.

In December 1999, even the White House had changed its position. New cryptography regulations were released, allowing for a wide variety of "automatic exemptions" from export restrictions. Subsequent tinkering led to an even more liberal policy: with a few exceptions, even the strongest cryptography could be exported directly overseas by U.S. companies.

The pressure exerted by news of RSA's DES Challenges might well have been just the force needed to cause the Clinton administration to reverse its position and to stop fighting industry efforts to address the global marketplace.

Today, software is different from what it was in 1997. Now, products come with strong cryptography built in. From both the perspective of forcing DES into retirement and allowing U.S. companies to participate in the global market for cryptography, the Crypto Wars—the battle to liberate cryptography—were won. While neither the DESCHALL group nor the RSA Secret Key Challenges can take sole credit, both are rightly seen as major contributors to one of the most critical battles.

While cryptography today is free in practice—through the absence of restrictions—it is noteworthy that the SAFE bill never did make it to the Senate floor, and its provisions prohibiting the government from introducing requirements for restricted cryptography never became law.

As a result of improved protection and reclaimed liberty, in 2004, many more people are accustomed to the idea of encryption and how it protects their information against threats to confidentiality and integrity. While they don't usually understand what exactly it means, they know the difference between "secure" and "unsecured" when they're shopping online. The infrastructure that supports electronic commerce and global communication is safer against a wide variety of attacks, and citizens are free to communicate privately with whomever they choose thanks to cryptography's accessibility.

We should not conclude that privacy is "solved" because we now have the freedom to encrypt.

The need to employ cryptography is becoming increasingly obvious, and not just for the purpose of transmission of information. Information stored in computers is now being encrypted with greater frequency. Even where information technology users have not historically been especially sophisticated (such as health care), industry regulation designed to improve the accountability of information handling now requires encryption of certain types of information under certain circumstances.

Cryptography is now also being used for more controversial purposes, and what appears to be the sequel to the Crypto Wars is already underway. This time, it isn't the balance of rights between individuals and their government, but consumers and the vendors who sell them things.

Consider the use of cryptography in new media, such as DVD players. Most DVDs produced are now encrypted, such that the discs will not play on devices that have not licensed the key needed to unlock the video stream. The idea is that by having discs secured against playback except on devices where the manufacturers have agreed to pay a fee to disc producers and to enforce certain rules, an exclusive club can be established for the playback of copyrighted work. Film producers expect that the exclusive club would effectively protect against the production of illegally copied discs.

Of course, cryptography is not a tool that prevents the copying of data. Cryptography is a tool that makes data—even if copied—useless to anyone without the key. Thus, cryptography's use in DVD players does nothing to prevent DVDs from being copied. Cryptography in

DVD players only provides an artificial extension of producer's rights to playback devices. It's a little like selling you a copy of this book but having the text be impossible to read unless you also buy a pair of glasses with a special chip that only I can sell. You can buy your glasses from anyone you choose, produced by any manufacturer you like, but the critical component needed to read the book you've already purchased must always be supplied by me.

That might sound like an anti-consumer position, and lawyers can (and have been) debating about the legalities involved. In addition to the law that has been historically applied in cases like this, the Digital Millennium Copyright Act (DMCA) is being argued. DMCA, which became effective in October 2000, updated U.S. copyright law, strengthening it considerably in favor of copyright holders.

In particular, DMCA prohibits any *attempt* to defeat an "effective" technical means of copyright enforcement. Putting the obvious logical question aside—an effective mechanism would withstand attack, so what's the point of prohibiting attack?—we are still left with a troubling question. If consumers cannot independently verify the security of such systems and if we cannot understand how these systems are likely to fail, how are we supposed to ensure their validity? Do we naively assume that "someone else" has taken care of it?

When faced with a technology that claims to protect publishers' rights without infringing consumers' rights, should consumers and copyright holders simply accept such claims at face value? Why would such claims not be subject to the same kind of public dissection and commentary that affect other rights, as was the case with cryptography?

Princeton professor Edward Felten led a team that responded to a challenge to crack technologies under consideration for the protection of digital information put forth by an industry group, the Secure Digital Music Initiative (SDMI). Felton's paper describing his inquiry and findings was scheduled for publication in a scientific context, at the Fourth International Data Hiding Workshop in April 2001. Upon learning that Felton's paper was to be published, SDMI and the Recording Industry Association of America (RIAA) threatened to sue Felton for violating DMCA because Felten's analysis of their digital "watermarking" methods showed how they could be defeated. After some threats of litigation were dropped and others resulted in suits being filed, the paper was finally published in August 2001 at the USENIX Security Symposium.[21]

(Other researchers have chosen to censor themselves rather than face threats of litigation by large industry cartels.[22])

Imagine a system designed to track the activity of Web users surreptitiously, employing cryptographic mechanisms to hide its activity—many of these systems have been discovered and documented.[23] If the user of a system wants to see what's happening, would he simply have to take the software manufacturer's word at face value? Would a manufacturer attempt to use DMCA to prevent analysis and commentary on technology that impacts the lives of its users?

In his book *Code and Other Laws of Cyberspace*, Lawrence Lessig makes a compelling argument that the technology all around us, the basis of our information infrastructure, is not inherently resistant to centralized control. Among the forces affecting the way that these systems work is the law. Because law also affects other forces, such as the market, it has a disproportionate influence.

As a consequence of DMCA, there is a body of law granting rights to copyright holders over how consumers may use their own devices, that they may not use them in such a way that mechanisms to protect the content are subverted. Indeed, part of that "protection mechanism" could involve having playback devices "phone home" to report user activity to the vendor.

As a result of the Crypto Wars, there is now largely an absence of law regarding the government's control of cryptography; citizens may use cryptography to communicate without government inspection. Copyright owners may also use cryptography to prevent consumers from seeing what playback devices are reporting when "phoning home."

Hence, increasingly strong publishers' rights in combination with the freedom of cryptography can present a danger to consumers. Simson Garfinkel's *Database Nation* covers the topic of privacy more generally, but one important point should be made here: privacy is not a "solved problem" because we are free to use cryptography. The people and organizations who want to watch our actions, whether for profit, to do us harm, or simply to get a cheap thrill are also free to encrypt.

Efforts to liberate cryptography have succeeded, and the world is now different as a result. In many ways, we're safer. In other ways, we're not. What is important to understand is that technology is amoral; it is neither good nor bad. Only people—free moral agents—can act to good or bad effect. Whether the freedom to encrypt helps us or hurts us ultimately depends on what we do with that freedom.

Notes

[1]Cipher Deavours and Louis Kruh. The Commercial Enigma: Beginnings of Machine Cryptography. *Cryptologia*, 26(1), January 2002.

[2]Jennifer Wilcox. Sharing the Burden: Women in Cryptology during World War II. NSA Web Site, March 1998. [online] http://www.nsa.gov/publications/publi00014.cfm.

[3]Stephen Budiansky. *Battle of Wits*. Free Press, 2002.

[4]G Johnson. Claude Shannon, Mathematician, Dies at 84. *The New York Times*, February 27, 2001.

[5]Claude E. Shannon. A Mathematical Theory of Communication. *Bell System Technical Journal.* 27:379-423 and 623-656, July and October 1948.

[6]Claude E. Shannon. Communication Theory of Secrecy Systems. *Bell System Technical Journal.* 28:656-715, October 1949

[7]Tom Athanasiou. DES and NSA's New Codes. In Peter G. Neumann, editor, *RISKS Digest,* volume 6, January 1987.

[8]Simon Singh. *The Code Book.* Anchor, 1999.

[9]D. Kahn. *The Codebreakers: The Story of Secret Writing.* Macmillan Publishing Company, New York, USA, 1967

[10]Robert Morris. The Data Encryption Standard—Retrospective and Prospects. *IEEE Communications Society Magazine*, 16(6):11–14, November 1978.

[11]National Bureau of Standards. Data Encryption Standard. Federal Information Processing Standards Pub. 46, Washington, D.C., Jan. 1977.

[12]Ruth M. Davis. Data Encyption Standard in Perspective. *IEEE Communications Society*, November 1978.

[13]Hayden B. Peake. The VENONA Progeny. *Naval War College Review,* 53(3), Summer 2000.

[14]Steven Levy. *Crypto.* Viking, 2001.

[15]Technically speaking, searching the keyspace would not take longer, but there would be more post-production work required to separate a possible match from a correct match. The difference, in practice, is negligible.

[16]Germano Caronni and Matt Robshaw. "How Exhausting is Exhaustive Search?" *CryptoBytes* 2(3), Winter 1997.

[17]The DESCHALL mailing list archives are still available online at http://www.interhack.net/projects/deschall/.

[18]András Salamon. Internet Statistics. [online] http://www.dns.net/andras/stats.html, February 1998.

[19]The first edition is online at http://www.crypto.com/papers/.

[20]The official abbreviation, which appears in RSA's documentation specifies more detail about the exact configuration of RC5 than just the key size. RSA wrote the fifty-six-bit version of RC5 as "RC5-32/12/7," which specified the "word size" (thirty-two bits), the number of "rounds" (twelve) the cipher would use, and the number of bytes for the key (seven, times eight bits for each byte gives us fifty-six bits).

[21]Information on the controversy and the paper itself can be downloaded from Princeton at http://www.princeton.edu/sip/sdmi/.

[22]Cryptographer Niels Ferguson has an essay on this topic, "Censorship in action: Why I don't publish my HDCP results." It can be found online at http://www.macfergus.com/niels/dmca/cia.html.

[23]One such system, PC Friendly, comes standard on many DVDs. See http://www.interhack.net/pubs/pcfriendly/.

Index